"This relevant and timely book situates research and pra ⟨…⟩ childhood education. Its international context demonstra ⟨…⟩ orative study and writing. The mathematical content is structured by using Bishop's six universal activities. Connecting these mathematical ideas with classroom activities offers the reader a novel perspective on both early childhood research and practice."

– *Alan J. Bishop, Emeritus Professor, Monash University, Australia*

MATHEMATICS IN EARLY CHILDHOOD

Structured around Bishop's six fundamental mathematical activities, this book brings together examples of mathematics education from a range of countries to help readers broaden their view on maths and its interrelationship to other aspects of life.

Considering different educational traditions and diverse contexts, and illustrating theory through the use of real-life vignettes throughout, this book encourages readers to review, reflect on, and critique their own practice when conducting activities on explaining, counting, measuring, locating, designing, and playing.

Aimed at early childhood educators and practitioners looking to improve the mathematics learning experience for all their students, this practical and accessible guide provides the knowledge and tools to help every child.

Oliver Thiel is Associate Professor at Queen Maud University College for Early Childhood Education, Norway.

Elena Severina is Associate Professor at Western Norway University of Applied Sciences, Norway.

Bob Perry is Emeritus Professor, Charles Sturt University, Australia.

EECERA
European Early Childhood
Education Research Association

TOWARDS AN ETHICAL PRAXIS
IN EARLY CHILDHOOD

Written in association with the European Early Childhood Education Research Association (EECERA), titles in this series will reflect the latest developments and most current research and practice in early childhood education on a global level. Feeding into and supporting the further development of the discipline as an exciting and urgent field of research and high academic endeavour, the series carries a particular focus on knowledge and reflection, which has huge relevance and topicality for those at the front line of decision making and professional practice.

Rather than following a linear approach of research to practice, this series offers a unique fusion of research, theoretical, conceptual and philosophical perspectives, values and ethics, and professional practice, which has been termed "Ethical Praxis."

Other titles published in association with the EECERA:

For more information about this series, please visit: www.routledge.com/education/series/EECERA

MATHEMATICS IN EARLY CHILDHOOD

Research, Reflexive Practice and Innovative Pedagogy

Edited by Oliver Thiel, Elena Severina, and Bob Perry

Routledge
Taylor & Francis Group

LONDON AND NEW YORK

First published 2021
by Routledge
2 Park Square, Milton Park, Abingdon, Oxon OX14 4RN

and by Routledge
52 Vanderbilt Avenue, New York, NY 10017

Routledge is an imprint of the Taylor & Francis Group, an informa business

British Library Cataloguing-in-Publication Data
A catalogue record for this book is available from the British Library

Library of Congress Cataloging-in-Publication Data
A catalog record has been requested for this book

ISBN: 978-0-367-37048-0 (hbk)
ISBN: 978-0-367-37050-3 (pbk)
ISBN: 978-0-429-35245-4 (ebk)

Typeset in Bembo
by Newgen Publishing UK

Printed in the United Kingdom
by Henry Ling Limited

CONTENTS

PART 3
Measuring 61

PART 5
Designing 133

PART 6
Playing 165

CONTRIBUTORS

Rosa María Vicente Álvarez, Doctor in Pedagogy, collaborates as a researcher in proposals on technology didactic materials and music education. An important contribution is the implementation and development of the "MusicChild" project directed by Dr Pieridou Skoutell. She participates in seminars and conferences about the evaluation of music materials and developing teaching around Europe. Her experience as a preschool, primary, and music teacher provides an extensive background for her work as a teacher at the University of Vigo. She organized the *International Symposium of Music Education and Didactic Materials* (Santiago de Compostela, 2019).

Sema Baydilli graduated in early childhood from Middle East Technical University. She has worked for eleven years: four years as an early childhood teacher; two years as an educational specialist; and five years as a principal. She has also participated in training programmes, guiding teacher education programmes, and administrative and management processes among the tasks required by the Ministry of Education as well as IB-PYP programmes with teachers. During her working experience, she also wrote early childhood mathematics books for children.

Oda Bjerknes is a kindergarten teacher with several years' experience teaching in traditional Norwegian kindergartens. Currently, she works as a maths and science teacher in Espira-parken, a facilitated maths and science centre for five-year-old children. In addition, she previously worked part-time with the Norwegian National Center for Mathematics in Education, assisting in the development of activities for the kindergarten department.

Elizabeth Carruthers is currently finishing her doctorate at the University of Bristol. She is the most recent head of Redcliffe Nursery School and National

Teaching School in Bristol, England. Her research interests are early mathematical graphics, teachers as researchers, and early years leadership. She has co-authored two books and writes for teachers and with teachers. Her work on mathematical graphics along with her co-author, Maulfry Worthington, was recognised by the English Department of Education, Williams Review of Early Years and Primary Mathematics.

Lucía Casal de la Fuente is a singer, psychopedagogue, and researcher at the Department of Pedagogy and Didactics of the University of Santiago de Compostela. Her training and expertise are focused on best practices in vocal education and singing. She has completed several international research stays to delve into singing therapy, childhood, and innovation and equity in education. She coordinates "Voces Ledas," a psychopedagogical project on vocal education and singing.

Alf Coles is Associate Professor of mathematics education at the University of Bristol, School of Education. He collaborates widely, and his research focus is on the early learning of number, the professional learning of teachers (with a particular interest in practitioner research) and, most recently, on how the mathematics curriculum (in school and higher education) can adapt to both the local concerns of communities (such as pollution) and to global issues (such as climate change). Alf currently co-edits the journal *Research in Mathematics Education*.

Sue Dockett is Emeritus Professor, Charles Sturt University, Australia. While recently retired from university life, Sue remains an active researcher in the field of early childhood education. Sue has been a long-time advocate for – and of – the importance of recognizing and responding to young children's perspectives. She maintains this position in her current work with children, families, and educators in explorations of transitions to school, children's play, and learning.

Simone Dunekacke is Professor at the Freie Universität Berlin, Germany, for Early Childhood Education Research. Her research interest is on early childhood teacher's professional competence, especially in the domain of early mathematics education. Her research addresses early childhood pre- and in-service teachers and focuses on action-related aspects of professional competence.

Maria Figueiredo is Associate Professor of Educational Sciences, Childhood Education, at the School of Education of the Polytechnic of Viseu, and a researcher at the CI&DEI/IPV. She has been Secretary-General of the European Educational Research Association since 2016. Her work is developed primarily in early childhood and primary school teacher education and social pedagogy. She has researched participatory pedagogies in early childhood education with a strong focus on children's rights and family involvement.

Carol Gillanders. Since the year 2000 Carol has taught music education at the Faculty of Education of the University of Santiago de Compostela. She is a member of the Educational Technology Research Group directed by Ms Beatriz Cebreiro and has worked in various European projects. Her research interests include service learning, interdisciplinary arts projects, and information and communications technology. She is co-author of a guide for teachers for accompanying songs as well as several articles related to music education.

Helena Gomes is Associate Professor of Mathematics at the School of Education of the Polytechnic of Viseu, and a researcher at the CIDMA/UA. She develops her research activity in the area of mathematics, specifically in algebraic graph theory, and about learning mathematics in early childhood education and in basic education. She has collaborated with the Ministry of Education in the preparation of support materials for the early years and in the evaluation and certification of mathematics textbooks for basic education.

Marie Hage was involved in early childhood education for 40 years as a teacher and director before retiring in 2018. She has been involved in numerous research projects around numeracy, the Reggio Emilia philosophy, and nature education. Marie values giving children uninterrupted time to interact with their environments, to hypothesise, observe, and critique their discoveries. Through observing and documenting children as they play, she has gained a deep understanding of how children best learn.

Elspeth Harley has worked in early childhood education since the late 1960s as a preschool teacher, university lecturer, curriculum adviser, and writer and research associate. She has a background in the arts and drama. Elspeth is committed to promoting the importance of play in the curriculum and in supporting educators to observe and document children's understandings of numeracy demonstrated in their play. Now retired, she mentors early childhood teachers in preschool and the first years of school settings and is a visiting storyteller and player at her local school and preschool.

Lone Hattingh is Senior Lecturer and Award Leader for MA Education: Early Childhood Studies at Bath Spa University. Her research interests include early literacy with a particular focus on the ways in which children engage with materials to make meaning. Lone is co-leader of REaCH, Bath Spa University's Centre of Research in Early Childhood, and is currently leading a project about perspectives of outdoor play which draws on case studies in England, Denmark, and the United States.

Trude Hoel is Associate Professor at the National Centre for Reading Education and Research at the University of Stavanger, Norway. She has taken part in several interdisciplinary projects in Norwegian kindergartens, about exciting things like woodlice and potatoes. Her research concerns early literacy, language learning, and children's book reading in print and on screen.

Børge Iversen is currently working as a preschool teacher in a municipal kindergarten in Trondheim, Norway. He has a master's degree in early childhood education and graduated as a pedagogista from the Reggio Emilia institute in Stockholm.

Aljoscha Jegodtka is currently teaching and researching at the IUBH – International University of Applied Sciences, Berlin. His research focuses on early mathematical education and the development of pedagogical professionalism among early childhood educators. At the same time, he carries out advanced training for early childhood teachers in the field of early mathematical education.

Kelly Johnston is a lecturer with the Department of Educational Studies at Macquarie University. She teaches across undergraduate and postgraduate units, specialising in mathematics, science, and technology in early childhood and primary school. Previously, Kelly was engaged in a wide range of roles across the early childhood education sector. This included working as an early childhood teacher and service director in Australia and New Zealand, teaching primary school in the UK, and working in early childhood licensing and accreditation at both a state and federal level in Australia. Kelly's research focuses on: exploring technology and digital citizenship for young children; mathematical language, thinking, and learning for infants and toddlers; and young children's learning in museum spaces.

Sonya Joseph has worked in early years education for 26 years. She has a bachelor of teaching – junior primary, and is a co-coordinator and educator at Faith Lutheran College, Early Learning Centre in the Barossa Valley. She is interested in the interconnectedness of relationship, environment, and learning and aspires to the principles of Reggio Emilia.

Camilla Normann Justnes is Assistant Professor at the Norwegian Centre for Mathematics Education, Norwegian University of Science and Education, Norway. Her research and development work focuses on mathematical discussions and inclusion in kindergarten and lower primary school. Justnes works foremost with developing resources for professional development in mathematics with and for kindergarten teachers, with a particular interest in building a positive view of mathematics for children and their teachers.

Ahmet Sami Konca is Assistant Professor at Erciyes University, Turkey. He has a bachelor's degree in elementary mathematics education, and a master's degree and a PhD in early childhood education. His main interests are early mathematics education; digital technologies in early childhood education, both in home settings and in classroom settings. He focuses on children's interactions with digital technologies, digital play, and technology integration into early mathematics education.

Andreas Lade is a young teacher in a Norwegian kindergarten with a bachelor's in language, texts and mathematics from the Western Norway University of Applied Sciences. He was one of three kindergarten student teachers who participated in the research project "Photography and image processing in kindergarten as a stimulating activity for mathematics learning", which directly led to his involvement in this book writing process. He used parts of the research project to further his bachelor's thesis on a similar yet different theme: "How does the kindergarten use the Technological Pedagogical Content Knowledge model (TPCK) to map the staff's knowledge on information and communication technologies (ICT)? He's currently employed as a kindergarten teacher at a private kindergarten in Bergen, Norway. He strives to lay the best path for the children's current and future education.

Myriam Marchese has a master's degree in early childhood and primary education. During her studies, she researched how children gave meaning to measurement in an early childhood setting. For three years, she has been working as an early childhood education teacher in the Associação de Beneficência Luso Alemã (ABLA) in Lisbon.

Ana Patrícia Martins is Associate Professor of mathematics at the School of Education of the Polytechnic of Viseu and a researcher at the CI&DEI/IPV and at the CIUHCT/FCUL. She holds a PhD in history and philosophy of sciences, a master's degree in mathematics teaching, and she graduated in mathematics. Her research interests include didactic of mathematics, history of mathematics, and history of science.

Luís Menezes is Professor at the Higher School of Education of Viseu, Portugal. His main research interests are mathematics teaching practices. In particular, he became interested in the role of communication in mathematics learning. Over 30 years of teaching and research, he has collaborated with the Ministry of Education, having participated in the design of national programmes of mathematics for basic education. In the last decade, he has a growing interest in childhood mathematics education.

Anne Hj. Nakken works both as an assistant professor at The Norwegian Centre for Mathematics Education, and as a preschool teacher. Her interests include developing, using, and evaluating research-based resources for preschool teachers, educational technology, transition from preschool to school, mathematics curriculum, and professional development. Anne has over the past 15 years worked on various research and implementation projects focused on improving preschool mathematics teaching and learning.

Zoi Nikiforidou is a senior lecturer in early childhood at Liverpool Hope University. Her research interests relate to aspects of cognition, pedagogy, and risk.

Zoi is a member of the OMEP UK executive committee and a co-convenor of the EECERA Special Interest Group *Holistic Wellbeing*.

Bob Perry has recently retired after 45 years of university teaching and research. He is Emeritus Professor at Charles Sturt University, Australia and Director, Peridot Education Pty Ltd. In conjunction with Sue Dockett, he continues his research, consultancy, and publication in early childhood mathematics education; educational transitions, with particular emphasis on transition to primary school; researching with children; and evaluation of educational programmes. Bob is co-chair of the EECERA Special Interest Group *Mathematics 0–8 Years*.

Elin Reikerås is Professor in early childhood education and Centre leader of FILIORUM – Centre for Research in Early Childhood Education, at the University of Stavanger, Norway. She has completed research on young children's mathematical development over the last 30 years and published a range of articles and books on the theme. She also led the interdisciplinary longitudinal Stavanger project "The Learning Child," which followed over 1,000 children's development from when they were two and a half years old until they turned ten years old.

António Ribeiro is Associate Professor of mathematics at the School of Education of the Polytechnic Institute of Viseu. He is a member of CI&DEI/IPV and develops his research activity in the area of didactics of mathematics with a particular interest in the role of teaching resources in the process of teaching and learning mathematics in the first and second cycles of basic education.

Janet Rose is currently principal of Norland College, a specialist early years higher education institution. Prior to this, she worked at several universities leading undergraduate and postgraduate early years degree programmes, becoming an associate professor and reader in education at Bath Spa University. She also worked with Elizabeth Carruthers in developing a postgraduate certificate in early years mathematics for specialist leaders to support early years mathematics in order to raise achievement in this subject.

Lilith Schechner studied part-time early childhood inclusive education (BA) at the Fulda University of Applied Sciences and worked at the "Kinderhort Gropiusstadt" in Berlin, where she is still employed as a state-recognized childhood educator and childhood educator for language.

Elena Severina is Associate Professor at Western Norway University of Applied Sciences with a PhD in physics and mathematics, and a postgraduate certificate in education. Elena teaches mathematics education courses to pre-service preschool, primary, and secondary school teachers and supervises bachelor's and master's students. Her research interests are in early childhood mathematics, and include critical thinking and argumentation, embodied cognition, spatial reasoning, child-centred digital photography activities, ICT, and multilingual classrooms.

Isabel Soares has been an early childhood teacher for over 35 years in Viseu. For many years, she has welcomed student teachers in her centre, supporting their practicum. She currently works in the Agrupamento de Escolas Infante D. Henrique. In her long professional career with children, she has always valued sensibility and respect as central dimensions of her practice.

Oliver Thiel is Associate Professor at Queen Maud University College in Trondheim, Norway. He has taught early childhood mathematics education in Germany and Norway for over 20 years. His research interests are teachers' attitudes and beliefs about mathematics and children's mathematical competence. Oliver is convenor of the EECERA Special Interest Group *Mathematics Birth to Eight Years*. He is editor of the EECERJ Special Issue *"Innovative Approaches in Early Childhood Mathematics"* and the German teacher journal *Mathematik differenziert* (Mathematics differentiated).

Maulfry Worthington taught for almost 30 years in the three- to seven-year age range, and has lectured in early childhood education, and primary and early years mathematics at the undergraduate and master's level in England. Her research interests include semiotics, graphicacy, children's mathematics, cultural learning, language acquisition, and pretend play. Her PhD (VU University, Amsterdam) investigates the natural history of signs in young children's mathematical inscriptions. Maulfry has published extensively, both individually and with Elizabeth Carruthers, and together they founded the international *Children's Mathematics Network*.

Lynne Zhang is an early childhood educator, researcher, and founder of Happy Future Connections Children and Family Research Centre which is a prestigious early childhood research, consultancy, and training institute in the social and cultural context for practitioners, parents, researchers, and policymakers in China. She is chief editor of the *China Early Childhood Home & Community Learning Report* and Playful Mathematics books, and developed early childhood mathematical courses, activities, and materials based on Chinese traditional culture. Lynne is Country Coordinator (China) of EECERA.

FOREWORD

Introduction to the EECERA Book Series

Tony Bertram and Chris Pascal

Underpinning aspirations

This timely, scholarly, and highly readable book edited by Oliver Thiel, Elena Severina, and Bob Perry entitled *Mathematics in Early Childhood* provides the eighth book in an innovative book series generated by the European Early Childhood Education Research Association (EECERA). The EECERA Book Series entitled *Towards an Ethical Praxis in Early Childhood*, offers an innovative and exemplary vehicle for the international early childhood sector to develop transformative pedagogy which demonstrates effective integrated praxis. The EECERA Book Series is designed to complement and link with the *European Early Childhood Education Research Journal* (EECERJ), which is primarily a worldwide academic platform for publishing research according to the highest international standards of scholarship. The EECERA Book Series aims to highlight pedagogic praxis in order to demonstrate how this knowledge can be used to develop and improve the quality of early education and care services to young children and their families.

Pedagogic approach

The approach taken in the book series will not be a linear one, but rather a praxeological one focused on praxis, meaning a focus on pedagogic action impregnated in theory and supported by a belief system. It is this fusion of practice, theoretical perspectives, ethics, and research which we term "Ethical Praxis." This fusion is embodied in all EECERA research and development activity, but we anticipate the book series will have a stronger focus on the development of pedagogic praxis and policy. In addition to offering a forum for plural, integrated pedagogic praxis, the series will offer a strong model of praxeological processes that will secure deep improvements in the educational experience of children and families, of professionals and researchers across international early childhood services.

The book series acknowledges pedagogy as a branch of professional/practical knowledge which is constructed in situated action in dialogue with theories and research and with beliefs (values and principles). Pedagogy is seen as an "ambiguous" space, not of one-between-two (theory and practice) but as one-between-three (actions, theories, and beliefs) in an interactive, constantly renewed triangulation. Convening beliefs, values, and principles; analysing practices; and dialoguing with several branches of knowledge (philosophy, history, anthropology, psychology, sociology, amongst others) constitutes the triangular movement of the creation of pedagogy. Pedagogy is thus based on praxis, in other words, action based on theory and sustained by belief systems. Contrary to other branches of knowledge which are identified by the definition of areas with well-defined frontiers, the pedagogical branch of knowledge is created in the ambiguity of a space which is aware of the frontiers but does not delimit them because its essence is in integration.

Praxeological intentions

There is a growing body of practitioner and practice-focused research which is reflected in the push at national and international levels to integrate research and analysis skills into the professional skill set of all early childhood practitioners. This is a reflection of the growing professionalism of the early childhood sector and its increased status internationally. The development of higher-order professional standards and increased accountability are reflective of these international trends as the status and importance of early education in the success of educational systems is acknowledged.

Each book in the series is designed to have the following praxeological features:

- strongly and transparently positioned in the sociocultural context of the authors
- practice or policy in dialogue with research, ethics, and with conceptual/theoretical perspectives
- topical and timely, focusing on key issues and new knowledge
- provocative, groundbreaking, innovative
- critical, dialogic, reflexive
- Eurocentric, giving voice to Europe's traditions and innovations but open to global contributions
- open, polyphonic, prismatic
- plural, multidisciplinary, multimethod
- praxeological, with a concern for power, values, and ethics; praxis; and a focus on action research, the learning community, and reflexive practitioners
- views early childhood pedagogy as a field in itself, not as applied psychology
- concerned with social justice, equity, diversity, and transformation

- concerned with professionalism and quality improvement
- working for a social science of the social
- NOT designed as a textbook for practice but as a text for professional and practice/policy development

This eighth book in our praxis series exemplifies these underpinning philosophies, pedagogical ethics, and scholarly intentions beautifully. We believe it is topical and much needed for these challenging times when educational practices for young children are coming under increasing pressures, focusing on key issues and new knowledge in the teaching and learning of early mathematics. The book is also provocative and critical, encouraging and opening polyphonic and multinational dialogues about our thinking and actions in developing high-quality early childhood services internationally.

1

REFLEXIVITY AND EARLY CHILDHOOD MATHEMATICS EDUCATION

Applying Bishop's universality to vignettes of young children's learning

Bob Perry, Oliver Thiel, and Elena Severina

Introduction

As we move into a new decade, it is clear that mathematics, and, in particular, early childhood mathematics, is in the international spotlight. The recent release of the 2018 Programme for International Student Assessment (PISA) (OECD, 2019) has forced many countries to reconsider their curriculum, learning, and teaching approaches as results have either stagnated or declined. There are now many studies which seem to show that early childhood mathematics achievement is a strong predictor of success in future school mathematics, other school subjects, and life itself (Carmichael, MacDonald, & McFarland-Piazza, 2014; Claessens & Engel, 2013; Duncan et al., 2007; Geary et al., 2013). As a result, across the globe, there has been much encouragement for early childhood professionals in both prior-to-school and school settings to engage with their children in mathematics learning, with one aim being to ensure that the children's standards of achievement are higher by the time they meet the first national or international assessment of their careers. However, a recent paper (Watts et al., 2018) has suggested that some of the earlier estimates of the impact of early mathematics interventions on later school mathematics success need to be treated with caution and may have overstated the case.

> Taken together, these results lead us to make two primary conclusions. First, correlational approaches to questions regarding longitudinal achievement patterns should be approached with great caution. Second, early learning does not appear to be an "inoculation" that necessarily produces later achievement gains, and consequently, theories regarding skill-building processes probably require some amount of revision.
>
> *(Watts et al., 2018, pp. 550–551)*

While this later evidence does not deny the importance of early childhood mathematics education for future achievement, it does suggest also that the value mathematics has for children in the present and how children might experience mathematics in their early childhood years be considered. In order to achieve this, early childhood educators, researchers, and policymakers are urged not only to reflect on their practice but also introduce the notion of reflexivity to their thinking. For many, however, the distinction between "reflection" and "reflexivity" is unclear. Bolton (2010, pp. 13–14) helps with two descriptions.

> *Reflection* is learning and developing through examining what we think happened on any occasion, and how we think others perceived the event and us, opening our practice to scrutiny by others, and studying data and texts from the wider sphere.
>
> ...
>
> *Reflexivity* is finding strategies to question our own attitudes, thought processes, values, assumptions, prejudices and habitual actions, to strive to understand our complex roles in relation to others. To be reflexive is to examine, for example, how we – seemingly unwittingly – are involved in creating social or professional structures counter to our own values. (Italics in original)

Early childhood educators are urged to move beyond reflection and towards reflexivity; to consider what they and others believe and value; and to undertake their practice on the basis of such reflexivity. They should ask not only "What happened?" but also "Why did it happen?" and "What can I do about it?" Such a move is particularly required in the area of early childhood mathematics education, which has often been minimised in early childhood settings in spite of its long history through luminaries such as Fröbel (Fröbel & Lilley, 1967) and Montessori (1912).

A strong influence on early childhood mathematics education over recent years is the advent of the neoliberal political and advocacy juggernaut known as STEM (Science, Technology, Engineering, and Mathematics). While the STEM movement has made mathematics visible, there is a danger that mathematics will be seen only to be the "servant" of science, technology, and engineering and that all mathematics will need to be drawn from these other disciplines or apply to them. Such an approach does fit well with early childhood approaches such as relevance, experiential learning, and play, but also has the potential to reduce realisation of the uniqueness of mathematics, particularly mathematical thinking, with a nature and approach which demand respect in its own right (Devlin, 2012; Hardy, 1940). In early childhood, mathematics provides opportunities for challenge, investigation, discovery, and sustained shared thinking (Siraj-Blatchford, 2007) that are not restricted to utilitarian applications, but also stimulate creative and innovative thinking in both young children and their educators (Shen & Edwards, 2017). Experiences with mathematics in the early years develop thinking and reasoning for young children's present and future (Katz, 2010).

Current research perspectives indicate that mathematics is important in the here and now of early childhood as well as into the future; that there are some consequences of early childhood mathematics education for later learning, although the scope of these is under question; that early childhood educators are urged to adopt reflexive practices; and that while mathematics is very important in young children's lives, it also has a role in developing particular forms of knowledge and thinking in its own right, both for the present and for the future. How can all of this be achieved in early childhood education settings in the best possible way for the child? One response has been what Moss (2014) has dubbed "schoolification" and which he has critiqued in the following way:

> "schoolification," an expressive term for primary schooling taking over early childhood institutions in a colonising manner (OECD, 2006, p. 62), leading to a school-like approach to the organisation of early childhood provision, the adoption of "the content and methods of the primary school" with a "detrimental effect on young children's learning" (OECD, 2001, p. 129), and "neglect of other important areas of early learning and development" (p. 42). While mathematics, language and science matter, the question is how best to work with them in early childhood education; while the problem is how to avoid them contributing to further schoolification by the spread of crude and oversimplified educational approaches that are at odds with the learning strategies of young children and that end up doing more harm than good.
>
> *(p. 37)*

In the field of early childhood mathematics education, this book seeks to help answer the call for approaches which work well but do not bring associated detrimental effects.

Genesis and purpose of this book

This book is the product of the Special Interest Group (SIG) on *Mathematics Birth to Eight Years* within the *European Early Childhood Education Research Association* (EECERA). (Information on both EECERA and the SIG can be obtained from www.eecera.org/). The EECERA SIG *Mathematics Birth to Eight Years*

> aims to coordinate and disseminate international research on the discourse in the emerging early childhood mathematics education field. It creates a space for shared thinking and for creating synergies between participants from a wide range of professional and scientific contexts to encourage a clearer articulation and understanding of early childhood pedagogy, policy and practice in relation to mathematics.
>
> *(EECERA, 2020)*

In 2017, at the EECERA conference in Bologna, Italy, SIG members determined that they would undertake a single research project across numerous jurisdictions, the results of which could be used to generate a book in the EECERA series *Towards an Ethical Praxis in Early Childhood*. After a nomination, review, and selection process, teams of early childhood researchers and educators generated vignettes about young children's mathematical experiences. These vignettes described real, rather than hypothetical, snapshots of children's experiences and provided rich examples of mathematics in early childhood settings. They were jointly developed by at least one early childhood researcher and at least one early childhood educator, bringing research and practice together. For each of these vignettes, three commentaries were written – the first by one of the researchers involved with the vignette; the second by an early childhood educator involved in writing the vignette; and the third by an early childhood mathematics education researcher working in a jurisdiction different from the origin of the vignette. Each chapter in the book consists of a vignette and the three associated commentaries. A follow-up face-to-face meeting of many of the authors was held at the EECERA conference in Thessaloniki in 2019 where decisions were made about the final content and nature of the book, and presentations were made about the overall research project.

The final product of the project focuses on children's mathematics learning through play and other pedagogical approaches and is based on research, policy, and practice from several different contexts around the world. It is a fusion of research and practice emanating from a consistent approach to analysis and critiquing vignettes of practice which encourages both researchers and practitioners to work together to consider their own practice and values associated with this practice reflexively. The book has met the overall objective of the project: to present innovative, research-based, and practical pedagogic praxis with the aim of enhancing young children's experiences in, and learning of, mathematics.

Theoretical basis for the book

The research project and each chapter of this book are based on an analysis by Alan Bishop (1988a, 1988b) of mathematics as a cultural pursuit, and his theory of mathematical enculturation. Bishop (1988b) has argued that there are six fundamental mathematical activities which "are both universal, in that they appear to be carried out by every cultural group ever studied, and also necessary and sufficient for the development of mathematical knowledge" (p. 182).

> Mathematics, as cultural knowledge, derives from humans engaging in these six universal activities in a sustained, and conscious manner. The activities can either be performed in a mutually exclusive way or, perhaps more significantly, by interacting together, as in "playing with numbers."
>
> *(Bishop, 1988b, p. 183)*

TABLE 1.1 Bishop's six fundamental mathematical activities (derived from Bishop, 1988b, pp. 182–183)

Fundamental mathematical activity	Bishop's descriptions
Counting	The use of a systematic way to compare and order discrete phenomena. It may involve tallying, or using objects or string to record, or special number words or names.
Locating	Exploring one's spatial environment and conceptualising and symbolising that environment, with models, diagrams, drawings, words, or other means.
Measuring	Quantifying qualities for the purposes of comparison and ordering, using objects or tokens as measuring devices with associated units or "measure-words."
Designing	Creating a shape or design for an object or for any part of one's spatial environment. It may involve making the object, as a "mental template," or symbolising it in some conventionalised way.
Playing	Devising, and engaging in, games and pastimes, with more or less formalised rules that all players must abide by.
Explaining	Finding ways to account for the existence of phenomena, be they religious, animistic, or scientific.

Each of the chapters in this book is assigned to one of Bishop's activities, but all of them highlight interactions among them. These activities and the explanations offered by Bishop (1988b) are listed in Table 1.1.

Structure of the book

There are twelve chapters in this book with a total of 37 authors (some of whom have more than one contribution) who are either working in and/or have cultural backgrounds in eleven countries. The first chapter – this chapter – has been written to provide the focus for the contributed chapters that follow. Chapters 2 to 11 contain vignettes and commentaries and are organised according to Bishop's fundamental mathematical activities.

The structure of each of Chapters 2 to 11 is the same, commencing with a vignette and followed by the three commentaries. Following the vignette and before the commentaries there is a set of Reflective Questions designed so that readers are urged to bring their own experiences, values, and beliefs to the vignette. These can then be compared with and critiqued against the commentaries. Each of the chapters is meant to be read as a whole, but separately from the other chapters. It is not anticipated that readers would read from front to back cover in one sweep. Rather, this is a book to be "dipped into" on the basis of a Bishop activity, interest in a particular topic, or perhaps on the cultural genesis of the authors. To this end, brief details are supplied in Table 1.2 for each of Chapters 2 to 11.

TABLE 1.2 Brief chapter details

Chapter title	Vignette topic	Key Bishop activity	Genesis of commentary authors		
			1	2	3
2. "One potato, two potatoes …" – Mathematics in an outdoor setting	This vignette shows how an outdoor activity, a potato harvest, gives rich possibilities for facilitating play-based mathematical activities adapted to children at different ages.	Explaining	Norway	Norway	Germany
3. "Even bigger than this world" – young children thinking about numbers through their mathematical graphics	In this vignette, the teacher has put out a range of maths equipment following on from children's previous enquiries about infinity. For example, there is a range of markers, clipboards, an abacus, metre rulers, and a number grid 1–100.	Counting	United Kingdom	United Kingdom	United Kingdom/South Africa
4. "Let's roll the dice" – Exploring amounts, counting, transcoding, and investigating invariance and variance	This vignette takes place in a German kindergarten. The children take turns throwing a dice, counting the number of pips on its faces, naming the respective numeral, and gathering that number of toy blocks.	Counting	Germany	Germany	Australia
5. Harry's Line Work	The vignette commences when Harry – a 4-year-old Australian preschool boy – seeks out Danette, the administrative officer for the preschool, who is at her desk. She is using a ruler to measure and Harry takes some interest in this, asking Danette what she is doing and how she is using the instrument.	Measuring	Australia	Australia	Norway/Germany

Chapter title	Vignette topic	Key Bishop activity	Genesis of commentary authors		
			1	2	3
6. "It's a lot of work" – A tailor's measuring tape in the dollhouse	There is a group of 26 children mainly from Roma families and low socio-economic background in a public early childhood centre in a small Portuguese city. During free playtime, a small group of children noticed a measuring tape that was left in the dollhouse with no explanation given.	Measuring	Portugal	Portugal	China
7. Fantastic Mr Fox	Three children aged 4 to 5 years play in the sandbox, and Emma, the kindergarten teacher, observes them. The children have listened to the book "The Fantastic Mr Fox" by Roald Dahl. They are afterwards occupied by foxes in their free play.	Locating	Norway	Norway	Germany
8. Ghost stairs and a ghost tree	The story comes from a project on photography and mathematics in a Norwegian kindergarten. Three groups of four 5-year-old children took pictures of what they found beautiful in the outdoor environment, selected photos, and used them to design a photobook on a computer with a touchscreen.	Locating	Norway/ Russia	Norway	United Kingdom/ Greece

(continued)

TABLE 1.2 (*Cont.*)

Chapter title	Vignette topic	Key Bishop activity	Genesis of commentary authors		
			1	*2*	*3*
9. Building bridges between maths and arts	A workshop for a single-room rural preschool in Spain with six children aged between 3 and 6 years was offered. The aim of the workshop was to link maths and music with the ultimate goal of promoting mathematical thinking in early childhood through music.	Designing	Spain	Spain	Norway
10. Geometry learning of children in digital activities	The vignette takes place in an early childhood setting in the suburb of a small Turkish town. The digital activity consisted of drawing a locomotive that was composed of triangles, rectangles, squares, and circles.	Designing	Turkey	Turkey	Norway/Russia
11. "This is the safe. It has a number and no one else knows it" – Playing with mathematics	The vignette focuses on an episode of pretend play in which the boys' collaborative dialogue grew from their personal interest in security safes. During the course of their play, the boys made reference to many aspects of number and quantities, culminating in communicating ideas through emergent mathematical inscriptions.	Playing	United Kingdom	United Kingdom	United Kingdom/ Denmark

The final chapter provides a synthesis of previous chapters, highlighting the connections between research and the professional practice of observing children, writing vignettes, and analysing them in terms of Bishop's fundamental activities. It uses the research data reported in each of the chapters to introduce an experiential learning framework which can be used to enhance young children's mathematical learning. The editors argue for the importance of reflexive practice as educators engage with, listen to, and respond to young children's early mathematical activities. The chapter concludes with a range of possible ways to continue research and praxis conversations around appropriate methods and methodologies for supporting early childhood mathematics teaching and learning.

The editors of *Mathematics in Early Childhood: Research, Reflexive Practice and Innovative Pedagogy* have enjoyed their role in creating this book and they commend it to anyone who is working in, or interested in, the field of early childhood mathematics education. It is a different book, bringing together a group of authors through a common objective: to enhance reflexivity in mathematics learning and teaching for young children. The book is innovative in design and rich in detail. Hopefully, it will stimulate early childhood educators and researchers to continue striving for the very best in mathematics education for their children and themselves.

References

Bishop, A. J. (1988a). *Mathematical enculturation: A cultural perspective on mathematics education.* Dordrecht: Kluwer.

Bishop, A. J. (1988b). Mathematics education in its cultural context. *Educational Studies in Mathematics, 19*, 179–191.

Bolton, G. (2010). *Reflective practice: Writing and professional development* (3rd ed.). Los Angeles: SAGE.

Carmichael, C., MacDonald, A., & McFarland-Piazza, L. (2014). Predictors of numeracy performance in national testing programs: Insights from the longitudinal study of Australian children. *British Educational Research Journal, 40*(4), 637–659.

Claessens, A., & Engel, M. (2013). How important is where you start? Early mathematics knowledge and later school success. *Teachers College Record, 115*(6). Retrieved from http://eric.ed.gov/?id=EJ1020177

Devlin, K. J. (2012). *Introduction to mathematical thinking.* Palo Alto: Keith Devlin.

Duncan, G. J., Dowsett, C. J., Claessens, A., Magnuson, K., & Huston, A. C. (2007). School readiness and later achievement. *Developmental Psychology, 43*(6), 1428–1446.

EECERA. (2020). *SIG Mathematics birth to eight years.* Retrieved from www.eecera.org/sigs/mathematics-birth-to-eight-years/

Fröbel, F., & Lilley, I. M. (1967). *Friedrich Froebel.* Cambridge: Cambridge University Press.

Geary, D. C., Hoard, M. K., Nugent, L., & Bailey, D. H. (2013). Adolescents' functional numeracy is predicted by their school entry number system knowledge. *PLOS ONE, 8*(1), e54651. Retrieved from PLOS website: http://journals.plos.org/plosone/article?id=10.1371/journal.pone.0054651

Hardy, G. H. (1940). *A mathematician's apology.* Cambridge: Cambridge University Press.

Katz, L. G. (2010). STEM in the early years. *Early Childhood Research and Practice.* Retrieved from ECRP website: http://ecrp.uiuc.edu/beyond/seed/katz.html

Montessori, M. (1912). *The Montessori method.* New York: Frederick. A. Stokes.

Moss, P. (2014). *Transformative change and real Utopias in early childhood education*. London: Routledge.

OECD. (2019). *PISA 2018 results*. Retrieved from www.oecd.org/pisa/publications/pisa-2018-results.htm

Shen, Y., & Edwards, C. P. (2017). Mathematical creativity for the youngest school children: Kindergarten to third grade teachers' interpretations of what it is and how to promote it. *The Mathematics Enthusiast, 14*(1), 325–345. Retrieved from University of Montana website: https://scholarworks.umt.edu/tme/vol14/iss1/19

Siraj-Blatchford, I. (2007). Creativity, communication and collaboration: The identification of pedagogic progression in sustained shared thinking. *Asia-Pacific Journal of Research in Early Childhood Education, 1*(2), 3–23.

Watts, T. W., Duncan, G. J., Clements, D. H., & Sarama, J. (2018). What is the long-run impact of learning mathematics during preschool? *Child Development, 89*(2), 539–555.

PART 1
Explaining

2

"ONE POTATO, TWO POTATOES ..."

Mathematics in an outdoor setting

Elin Reikerås, Trude Hoel, Børge Iversen, and
Aljoscha Jakob Jegodtka

Vignette

The aim of this vignette is to show how a two-hour-long outdoor activity, a potato harvest, gives rich possibilities to facilitate play-based mathematical activities adapted to children at different ages. The kindergarten where the potato harvest takes place is housed in an old villa in a central residential area in a Norwegian city. A large and old orchard with fruit trees, sandpits and swings, a playground with grass, and areas where the children grow herbs and vegetables surrounds the villa. In an old sandpit, there is a potato field. The children planted the potatoes in early May, and all summer they have tended the growing of the potato plants, giving them water and weeding the area. In September, it is time for potato harvest. The participants are 18 children aged three to six years old, nine children aged one to three years old, two kindergarten teachers, four assistants without formal teacher training and two researchers observing the activity.

This day the rain pours down. The children, the staff, and the researchers are dressed in their rain clothes as they gather around the potato field. The children have buckets and spades, and one of the teachers handles a pitch-fork. The children gather potatoes as they appear from the soil. They also gather other things that emerge, like earthworms. The staff take an active part in the distribution of potatoes so all the children get a little each of the somewhat meagre potato harvest. On a paved area outside the kindergarten's kitchen door, a basin of water is put out. The children carry the potatoes from the field in their buckets and put them into the water. All the children get to wash the potatoes. Thereupon the children classify the potatoes into categories according to different characteristics. The children sort the

FIGURE 2.1 Buckets labelled for categorising potatoes

potatoes according to size and shape and develop categories (Figure 2.1), such as "large," "small," "middle," "smallest," "rarest," and "most dangerous."

A group of four- to five-year-olds has written labels for the various categories. The staff have supported the writing, meaning the children wrote some of the labels themselves, and the staff penned others. The labels are used to mark small victory podiums made of empty tin cans (Figure 2.2).

The labels are laminated so they will endure the weather. The children negotiate which potatoes to place on which victory podium. The categories are not completely separated, and problem-solving and negotiation become necessary. Through their sorting, a potato is placed on the small-potato podium. Further on, a girl finds a potato about the same size and says, "Now we have two smallest." Then two children come forward with one more potato as small as the two others. The teacher asks if it is possible to have many smallest or only one. The three children agree that there can only be one "smallest." They study the potatoes carefully and negotiate additional categories, "smallest," "even smaller," and "very small." Only one potato can be the winner, and it is difficult to see the difference in size, so they agree to choose the cutest potato as the winner.

Potato harvest is a tradition in the kindergarten and some of the four- to five-year-old children remember from last year that they had made a row with all the potatoes and start lining up this year's harvest in a line ordered by size. Negotiations about the size and shape of the potatoes characterise this activity. When the line is finished, one of the girls is curious about whether the row is longer this year than last year. She and a boy lie down beside the

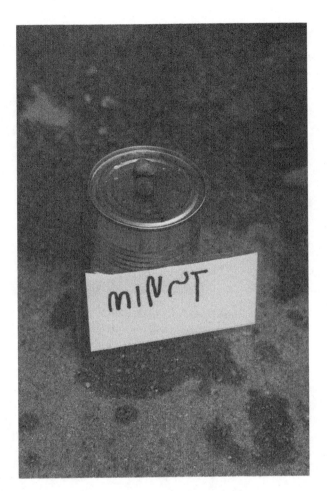

FIGURE 2.2 Podium for the smallest potato

row (Figure 2.3) and measure it to be nearly "two large children long." It was much longer last year when it was nearly "three large children long."

One of the four-year-old children begins to count the potatoes when they are in the row. She says the number sequence as long as she can, stopping at 27, without pointing at the potatoes. The teacher suggests she can count the large potatoes first and point at the potatoes while counting. Then it becomes much more difficult. The number sequence is difficult for her to recite when trying to point at the same time as saying the counting words, especially when the number sequence passes ten. Her four last counting words are eleven, twenty, seventeen, eighteen. "It is eighteen large potatoes," the teacher concludes. Then the girl counts the small potatoes, and this time she is not as careful with her pointing. She finalises her count with "twenty-nine, twenty-ten, twenty-eleven, twenty-twelve, twenty-thirteen potatoes!"

FIGURE 2.3 A row of potatoes and children

FIGURE 2.4 Children counting potatoes

"Therefore, thirty-three small potatoes," the teacher translates. Other children also join in counting (Figure 2.4). There are many different answers to how many potatoes they have. The teacher suggests they should count together while putting the potatoes into the bucket. In this counting, the teacher also counts together with the children, and they finally sum up 44 potatoes.

The potatoes finally end as a shared meal prepared by the children and staff together.

Reflective questions

1 What roles are adopted by the educators in the vignette? How do these contribute to the children's experiences and their mathematical competence?
2 Consider why the educator might have chosen not to correct the children's use of the counting sequence. What strategies are being used to prompt children's mathematical thinking?
3 What characteristics of the potatoes and the activity do you think were important to provoke the children's exploration and thinking?

Commentary 1

Authors: Elin Reikerås and Trude Hoel

In the Nordic countries, there is a culture of seeing children's outdoor play, especially in nature, as part of a good childhood (Waller et al., 2010). Thus, in Norway, early experiences with outdoor conditions are regarded as both important and healthy, not least in kindergartens (Norwegian early years' education and care institutions for children of age one to five years).

The outdoor environment is considered to be a pedagogical space and in the *Framework Plan for Kindergartens – Content and Tasks* (Ministry of Education and Research, 2017), being outdoors is emphasised as an important arena, especially for play. Norwegian children spend a significant amount of time outdoors in kindergarten: 70% in summer and 30% in winter (Moser & Martinsen, 2010). However, Norwegian kindergarten teachers do not recognise or fully use learning possibilities in outdoor settings (Kaarby & Tandberg, 2017).

There is an increasing body of international literature giving empirical evidence of the value of children's outdoor activities in terms of physical, health, and social benefits. However, research concerning outdoor space for cognitive aspects of development and learning is not often addressed, although outdoor activities provide other opportunities for learning than indoor activities do (Waters & Maynard, 2010). In addition, there has been minimal attention paid to the opportunities that outdoor spaces may offer in terms of supporting adult-child talk and the potential for associated cognitive engagement (Maynard, Waters, & Clement, 2013), such as in mathematics.

The Framework Plan underlines that all kindergartens must take a holistic approach to children's development, with care, play, and learning as core values (Ministry of Education and Research, 2017). This is also the case in mathematics, which is one of seven learning areas, called *quantities, spaces, and shapes*. Thus, kindergartens are following a social pedagogical tradition in which children's participation, their own play, and daily-life activities are seen as more important than formal teacher-driven activities (Jensen, 2009; OECD, 2006). This characterises how mathematics is implemented in the Framework Plan, emphasising the kindergarten's responsibility to encourage children in their own investigations and enabling them, for example, to

- discover and wonder at mathematical relationships;
- play and experiment with numbers, quantities, and counting and gain experience of different ways of expressing these; and
- gain experience of quantities in their surroundings and compare them (Ministry of Education and Research, 2017, p. 53).

The work with mathematics in Norwegian kindergartens is often classified in line with Bishop's six fundamental mathematical activities (Bishop, 1991), adapted to young children's culture and mathematical activities by Solem and Reikerås (2017), in line with the social pedagogical tradition.

Even though the social pedagogical approach – represented by Piaget's and Vygotsky's theories – are important for pedagogical work in kindergartens, a central approach in outdoor settings is Gibsonian (Sandseter, Little, & Wyver, 2012). The Gibsonian theory emphasises direct engagement with the environment as necessary to learn about the environment. This approach is appraised as bottom-up, whereas the social pedagogical perspectives are seen as top-down, considering the relationship between the environment and learning to be mediated by cognitive and linguistic processes (Sandseter et al., 2012).

The potato harvest presented in the vignette facilitates various mathematical activities within an outdoor setting. One of these activities is participation in problem-solving and negotiating during the sorting and categorising of potatoes. The children sorted the potatoes according to size and shape and developed categories based on ranking such as "large," "small," "middle," and "smallest." In addition, they used characteristics based on other features, "funniest," "cutest," and so on, when sorting. This categorisation included all the participating children regardless of age and language skills. The youngest children – needing to enrich their mathematical vocabulary – met and explored new words that characterise the shape and size of the potatoes. At the same time, the older children had opportunities to build conceptual depth as they generated new associative networks based on already established terms. In this activity, the children experience both direct engagement with the environment, as emphasised by the Gibsonian approach, and a socially engaging context where both children as well as adults explore sizes, categorise, and develop mathematical concepts, as emphasised by the social pedagogical approach.

The teacher uses mathematical terminology thoughtfully and actively, as underlined in the Framework Plan (Ministry of Education and Research, 2017 p. 54).

Measuring the row of potatoes and the children's own counting activities are parts of a larger context based on the children's wish to investigate "how long" and "how many." When measuring how long the potato row was, the children used their own bodies. Although this is inaccurate, the children can handle this measurement unit better than standard measures such as metres, which are complicated and not easy for four- to five-year-olds to master (Sarama & Clements, 2009). Here the teacher encourages the children to "be curious, find pleasure in mathematics and take an interest in mathematical relationships based on their forms of expression" (Ministry of Education and Research, 2017 p. 54).

The children's different answers for "How many potatoes?" are not surprising since the development of counting and enumeration are complex processes that develop through the entire childhood (Mix, Huttenlocher, & Levine, 2002), particularly in play. This is also the case with the four-year-old girl. She could recite the number sequence to 27, whereas she could not combine this with pairing one number word to each potato. Such combined sequencing is reported to be only partly mastered before the age of five years (Sarama & Clements, 2009).

The kindergarten teacher facilitates the counting activity and gives support to the children's exploring. She acts, for example, as a guide when she suggests pointing. Further, she shows respect for the girl's counting by confirming her answer instead of correcting. The teacher scaffolds the girl's learning, according to social pedagogical perspectives on learning (Rieber et al., 2004). The teacher let the children explore counting and play with numbers. Later, she supports learning by counting together with them. This is in line with the guidelines of the Framework Plan, where the staff shall:

• create opportunities for mathematical experiences by enriching the children's play and day-to-day lives with mathematical ideas and in-depth conversations;
• stimulate and support the children's capacity for and perseverance in problem-solving (Ministry of Education and Research, 2017 p. 54).

Children in kindergarten learn best in purposeful contexts, where the staff support the child in both informal and more structured activities (Frye, 2013). In addition, the staff should recognise the concepts and skills the children have, to be able to facilitate further learning. The potato harvest shows how mathematical activities can be directed to children's individual interests and level of skills and how the staff can scaffold the different activities.

Although, to a large extent, the children decide in which ways they involve themselves within the frames that are facilitated, the kindergarten staff have an important role in following up and supporting the children in their learning. The staff need to be flexible, improvise, and see the disciplinary and mathematical potential in situations that occur. The potato harvest is one example of an outdoor activity with mathematical learning potential, which can easily be adapted to other outdoor activities.

The vignette is a glimpse into a day full of play and different activities within the frame of a potato harvest, which involves all the children in the kindergarten. The staff initiate some of the activities, whereas the children are free to participate in different ways based on interests and age. For some children the washing of the potatoes was the activity they engaged in most. While a five-year-old boy wrote labels for classifications of potatoes, a three-year-old girl counted how many earthworms she had collected in her bucket. In the soil, there is something for every interest, although it is not necessarily potatoes, thus allowing the potato harvest to present individual experiential and learning opportunities. In addition, by not separating the children into age-specific groups, they are able to learn from and be inspired by each other, as when the children lined up the potatoes and measured with their own bodies – as they observed the older children doing last year. The potato harvest is an example of how solving mathematical challenges together has a social purpose within a kindergarten context. Counting, classifying, and measuring are mathematical topics enhanced in the potato harvest activity. However, the activity does open up for engagement with other mathematical topics such as spatial sense, estimation, and patterns as well. The potato harvest includes examples of interdisciplinary learning activities such as science, language, and mathematics. Whether these activities are going to be mathematical activities depends, to a large extent, on staff competence and engagement.

Commentary 2

Author: Børge Iversen

This vignette showcases an outdoor activity where the teachers take advantage of the environment in which they are situated. According to Gibson (1979), we can say that an element of the environment, in this case the potatoes, has the potential of providing the children with some kind of function, or *affordance*. The potatoes become an artefact they can explore for mathematical, among other, purposes.

Without the obvious mathematical component, we also notice the activity is done jointly. It is a group effort, where the children engage in negotiations with each other. They are not presented with a blueprint on how the task is supposed to be solved. In other words, the children are making a concerted activity, which in itself creates the opportunity for experience within a specific discipline. Even though the main focus here is on mathematics, the approach can be said to be somewhat multidisciplinary. This corresponds with the philosophy of a project-oriented practice, inspired by the municipal preschools in Reggio Emilia (Malaguzzi, 1996). In such a practice, pedagogical documentation is an important tool, both for planning and professional reflection (Rinaldi, 2001). In this example, the vignette itself can function as pedagogical documentation, where many possible perspectives emerge.

It is also worth noticing how the potatoes serve as something in relation to which the children are allowed to assume a subject position. This corresponds to the thinking of the Norwegian philosopher Skjervheim, as something he calls *the third* (Sørbø, 2004). In short, this means the children can assume a subject position in relation to the potato and are not treated as objects themselves. The potato – or the third – is something outside of the subjects relating to it. This requires an object or a case the children can find meaningful. In this example, the potato becomes a third component, which provides the children with a shared focus, something they can explore jointly. This is consistent with the project philosophy, where the children are encouraged to find their own questions and to research them. Scientists work in much the same way. They formulate a problem and find a suitable method, from which they construct new and unique knowledge.

Even though a project is multidisciplinary, and the children might have more of a holistic approach to their surroundings, it might still be fruitful for the teachers to focus on a certain discipline, in this case mathematics, both to learn how to recognise the discipline in itself, and to understand it better in the context of everyday activities. Teachers who develop such competence will presumably be able to plan for better and more diverse mathematics activities. This will again benefit the children.

What can we then read from this vignette? The children are obviously engaged in harvesting potatoes, and it seems this is a project they have been committed to over a longer period. In this particular activity, it also seems the teachers have set the stage for the children to investigate some specific mathematical problems. This can be seen as the beginning of the project, where the overall theme is the potatoes, but we at this point chose to focus mainly on mathematics. This choice can be justified by referring to the documentation. It tells us the children are in need of a deeper understanding of some mathematical principles, to be able to continue their activity. It is not something the teachers force onto them. This is meaningful and useful to them, in this particular context. They probably do not think of it as learning mathematics in an educational sense. It is just an understanding and skill they need at that moment, to be able to continue their activity.

If we choose to see this through the lens of a project-oriented practice, it is helpful to view the vignette as documentation. This way it can become a tool for pedagogical reflection, where we can help expand the children's understanding and at the same time ensure their right to participate in the process.

In a project, the teachers may plan for different activities based on documentation of what the children have done previously. What questions have they asked? What have they tried figuring out? However, the children's direct participation will also be important in shaping and adapting the activity.

The teachers' role is also important during the activity itself. As we can learn from this example, they do not just passively observe the children. They ask questions and make suggestions. That way the teachers can encourage the children to experiment with different solutions to their problems. In the vignette, the children are encouraged to find creative solutions, in terms of making their own classifications, ordering, and ranking the potatoes. The teachers are staging situations where the children can work together with varied mathematical challenges.

So how could we envisage the continuation of this project? First, it would be a good idea to use the documentation together with the staff. That way we can share our different interpretations and reflect on what we see. This will both help the staff recognise the discipline in itself, and better understand how it is a natural part of the children's everyday activities. In this example, we can see at least three mathematical activities: measuring, counting, and explaining (Bishop, 1988). We see the children measure the potatoes by putting them in a row and lying down beside them to find that the row is shorter than last year. One of the children starts counting the potatoes when they are in a row. When they classify the potatoes by size, they are explaining similarities and differences by putting them into a hierarchy.

With the help of an inspiring physical environment, engaging teachers, and a richness of relevant artefacts, we can envision the children getting increasingly more aware of mathematical ideas. The teachers should always be on the lookout for moments where the children are engaged in mathematical activities. This could occur during meals, where the children can be challenged to put the right number of plates on the table; or in conversations related to a story or a book. This way the project can easily gain momentum and be guided by the children's own interests.

The teachers' job is to set the stage for teachable moments in creative environments, where the children potentially can meet and learn from each other. Since this particular interest started with the potatoes, it would be a good idea to take advantage of that in the continuation. When preparing a meal of the potatoes, the children could try to follow a recipe. In such a scenario they will need different mathematical skills, among others counting and measuring. This way we could both help the children gain a better understanding of the potato in itself and include the relevance of a mathematical perspective.

This way of thinking can be understood as a democratic and value-driven practice. In short, the children are given opportunities to work together, with challenges in line with their own interests. The teachers base their planning on their knowledge of the children's interests. This way we can ensure the children's voices are brought to the fore, while we simultaneously inspire them to explore the world of mathematics further.

Commentary 3: domain-specific learning in a play-based outdoor setting: a German perspective

Author: Aljoscha Jegodtka

The situation described in the vignette would rarely be found in this specific form in Germany. Although many kindergartens have a playground, putting together a vegetable patch is not very common. In Berlin, for example, there is no guarantee that the kindergartens will even have a playground at all. Often early childhood educators visit public playgrounds with their children as an outdoor activity. Such situations are thus more likely to occur in a specialised forest or nature kindergarten (Miklitz, 2011, 2019). According to the Federal Association of Nature and Forest Kindergartens in Germany (Bundesverband der Natur, 2020), there are just over 1,500 of these kindergartens – a small fraction of the approximately 55,000 kindergartens in Germany.

Project work in German kindergartens

However, the described methodological approach is widespread in Germany and called the "project approach," "project method," or "project work" (AG Projektarbeit, 2012; Jacobs, 2012; Reichert-Garschhammer, 2013; Stramer-Brandt, 2018).

> A project in educational institutions is an openly planned educational activity of a learning community of children and adults. The focus is on the intensive, interdisciplinary examination of a topic and its embedding in larger contexts. Together, the topic is studied and researched in a versatile and collaborative way; the questions and problems arising are identified and creative solutions are sought and developed together, which extends over a longer period of time.
>
> *(AG Projektarbeit, 2012, p. 6)*

The children learn independently and cooperatively, through researching and making discoveries, holistically and exemplarily. Projects are interdisciplinary and comprise several educational areas. The core of every project is the self-activity and personal responsibility of the child. Projects require step-by-step planning while maintaining flexibility in terms of goals set, the pursuit of these goals, and the design of daily routine, learning environment, and methodology. Times of activity and times of reflection alternate. The goals of the project can be adjusted over time. The final presentation and documentation of the project are important. Projects should not be special events. Rather, they emerge in everyday life. With projects, day-to-day routines are disrupted and redesigned.

A project usually consists of four phases:

Phase one: Project identification and clarification (initiation and entry phase). Decisive for the success of a project is the topic. As far as possible, it should interest all children and motivate them to do research. During this phase, children will discuss what they already know about the topic. This also results in a small subdivision of the topic into several aspects.

Phase two: Project planning. The plan is developed togethe[r]
dren. It defines what the first tasks are and what further
clarified who takes which task and which material is neede[d]
is scaffolding and should not determine the project. This
children in the project development right from the start. Th[e]
interact with others and work together on a topic.

Phase three: Project implementation. Depending on the nature of the project, the implementation is different. It is recommended not to overburden the project with activities. Each project involves joint exploration activities, but it is just as important that the children have enough freedom to explore individually and work on the respective tasks.

Phase four: Project completion and reflection. The presentation of the results of a project can happen over and over again. Nevertheless, a final presentation or closing activity should be part of it. Here a new project can be conceived through reflection on the past project.

Projects as a pedagogical approach can have different contents. Projects are characterised by the four phases. (see, for example, Haus der kleinen Forscher, 2020).

Early mathematics in German kindergartens

Early mathematical education in kindergartens in Germany should be integrated holistically and into everyday life. At the national level there is only a very rough framing of mathematics in the early childhood years. Thus, there is a lot of variation at the level of the federal states. Central are the mathematical guiding ideas or, in the sense of Bishop (1988), fundamental activities. According to German mathematical pedagogical research, teaching and learning build on these guiding ideas or fundamental activities. In addition, German mathematics education focuses on process-related competencies such as arguing and communicating (Benz et al., 2015). Likewise, the distinction between mathematics as pure science and mathematics as applied science plays a role. Children may want to explore the comparability of quantities with dice. This is a mathematical problem – mathematics as pure science. On the other hand, children can also ask themselves how to divide five pieces of candy among three children. Here mathematical solutions and mathematical knowledge help. This is a life-world problem – mathematics as applied science (Wittmann, 2016).

Taken together, this means that early mathematical education in kindergartens includes mathematics as pure science and mathematics as applied science, refers to guiding principles, enforces process-related competencies, and should take place holistically and be integrated into everyday life. This is a big challenge for educators.

Relation to the vignette

A project such as the described potato harvest featuring mathematical aspects can very well take place in German kindergartens – although it is unlikely that it is a real potato harvest. Similar projects involving both biological and mathematical aspects are well conceivable.

The children harvest the potatoes and they are divided among them so that each child has potatoes. These potatoes are then sorted according to the children's criteria. This is a genuinely mathematical activity, sorting, performed by the children. In particular, the criterion of size plays a role here. The children discuss how to sort the potatoes according to different criteria. This is the process-related competency of argumentation, which is part of the project and belongs to Bishop's (1988) fundamental activity "explaining." This is particularly evident when the children want to assign several potatoes to the position "smallest." Through the intervention of the early childhood teacher, it is discussed whether there may be one or many smallest potatoes.

The potatoes are ordered according to size and not according to other criteria. This creates a long row with the largest potato at one end and starting from that, smaller and smaller potatoes. The comparison of sizes is discussed by the children.

This is followed by a final mathematical sequence: naming the number words and counting something (Ginsburg, 1989). It is important to differentiate two moments of counting: counting in the sense of (correctly) saying the number words and counting in the sense of determining the number of elements of a set. A child wants to count the potatoes. Here the child names the known number words and stops at 27, but does not point to the potatoes. This shows a typical stage of development: the child already knows that the number words must be said to determine the number. However, she does not yet fully know the counting principles (see Jegodtka, Schechner, and Johnston, Chapter 4). The child is not yet aware of the one-to-one correspondence between the number word and the elements to be counted. In this sense, the teacher intervenes and shows the child, or later the children, counting strategies (Hess, 2012).

Another typical situation can be observed (Ginsburg, 1989): when children learn the order of the number words, they first discover one number word after another. Over time, they discover that the number words also have a fixed order, but there are many confusing ones. It is then a great step in children's development when they realise that the number sequence is structured according to principles. In this case, the child counts, "twenty-nine, twenty-ten, twenty-eleven" and so on. Here, the child gets help in recognising the right principles by the teacher, who is working as a language model, without criticising the child.

References

AG Projektarbeit (2012). *Gelingende BayBEP-Umsetzung in Kitas: Schlüssel Projektarbeit. Ein Positionspapier der AG Projektarbeit im Ko-Kita-Netzwerk Bayern* [Successful implementation of BayBEP in daycare centres: key project work]. A position paper of the Working Group Project Work in the Co-daycare-network]. Retrieved from www.ifp.bayern.de/imperia/md/content/stmas/ifp/ko-kita/gelingendebaybep-umsetzung_projektarbeitinkitas_positionspapier_stand_05-07-2012.pdf

Benz, C., Peter-Koop, A., & Grüßing, M. (2015). *Frühe mathematische Bildung. Mathematiklernen der Drei- bis Achtjährigen* [Early mathematical education. Mathematics learning of three-to-eight-year-olds]. Wiesbaden: Springer.

Bishop, A. J. (1988). Mathematics education in its cultural context. *Educational Studies in Mathematics, 19*(2), 179–191. Retrieved from https://www.jstor.org/stable/3482573.

Bishop, A. (1991). *Mathematical enculturation*. Dordrecht: Kluwer.

Bundesverband der Natur (2020). *Bundesverband der Natur – und Waldkindergarten in Deutschland* [Federal association of Nature and Forest Kindergartens in Germany]. Retrieved from https://bvnw.de/

Frye, D. (2013). *Teaching math to young children: a practice guide*. Washington, DC: National Center for Education Evaluation and Regional Assistance (NCEE), Institute of Education Sciences, U.S. Department of Education.

Gibson, J. J. (1979). *The ecological approach to visual perception*. Boston: Houghton Mifflin Harcourt.

Ginsburg, H. P. (1989). *Children's arithmetic: how they learn it and how you teach it*. Austin, Texas: Pro-Ed.

Haus der kleinen Forscher (2020). *Haus der kleinen Forscher* [House of little explorers]. Retrieved from www.haus-der-kleinen-forscher.de

Hess, K. (2012). *Kinder brauchen Strategien. Eine frühe Sicht auf mathematisches Verstehen* [Children need strategies. An early view on mathematical understanding]. Hannover: Friedrich Verlag.

Jacobs, D. (2012). *Projektarbeit: Kitaleben mit Kindern gestalten* [Project work: shape daycare life with children]. Weimar: Verlag das Netz.

Jensen, B. (2009). A Nordic approach to Early Childhood Education (ECE) and socially endangered children. *European Early Childhood Education Research Journal, 17*(1), 7–21.

Kaarby, K. M. E., & Tandberg, C. (2017). The belief in outdoor play and learning. *Journal of the European Teacher Education Network, 12*, 25–36.

Malaguzzi, L. (Ed.). (1996). *The hundred languages of children: narrative of the possible*. Reggio Emilia: Municipality of Reggio Emilia.

Maynard, T., Waters, J., & Clement, J. (2013). Moving outdoors: further explorations of "child initiated" learning in the outdoor environment. *Education Week, 41*(3), 282–299.

Miklitz, I. (2011). *Der Waldkindergarten: Dimensionen eines pädagogischen Ansatzes* [The forest kindergarten: dimensions of a pedagogical approach] (7th ed.). Berlin: Cornlesen.

Miklitz, I. (2019). *Naturraum-Pädagogik in der Kita. Pädagogische Ansätze auf einen Blick* [Nature space pedagogy in the daycare centre. Pedagogical approaches at a glance]. Freiburg i.Br.: Herder.

Ministry of Education and Research (2017). *Framework plan for kindergartens – content and tasks*. Oslo: Norwegian Directorate for Education and Training. Retrieved from www.udir.no/globalassets/filer/barnehage/rammeplan/framework-plan-for-kindergartens2-2017.pdf

Mix, K. S., Huttenlocher, J., & Levine, S. C. (2002). *Quantitative development in infancy and early childhood*. Oxford: Oxford University Press.

Moser, T., & Martinsen, M. T. (2010). The outdoor environment in Norwegian kindergartens as pedagogical space for toddlers' play, learning and development. *European Early Childhood Education Research Journal, 18*(4), 457–471. https://doi.org/10.1080/1350293X.2010.525931

OECD. (2006). *Starting Strong II: early childhood education and care*. Paris: OECD. Retrieved from www.oecd.org/dataoecd/14/32/37425999.pdf

Reichert-Garschhammer, E. (2013). *Projektarbeit im Fokus: Fachliche Standards und Praxisbeispiele für Kitas* [Project work in focus: professional standards and practice examples for daycare centres]. Berlin: Cornelsen.

Rieber, R. W., Robinson, D. K., Bruner, J. S., & Vygotskij, L. S. (2004). *The essential Vygotsky.* New York: Kluwer.

Rinaldi, C. (2001). Documentation and assessment: what is the relationship? In C. Giudici, C. Rinaldi, & M. Krechevsky (Eds.), *Making learning visible: children as individual and group learners* (pp. 78–89). Reggio Emilia: Reggio Children.

Sandseter, E. B. H., Little, H., & Wyver, S. (2012). Do theory and pedagogy have an impact on provisions for outdoor learning? A comparison of approaches in Australia and Norway. *Journal of Adventure Education and Outdoor Learning, 12*(3), 167–182.

Sarama, J. A., & Clements, D. H. (2009). *Early childhood mathematics education research: learning trajectories for young children.* New York: Routledge.

Solem, I. H., & Reikerås, E. (2017). *Det matematiske barnet* [The mathematical child] (3rd ed.). Bergen: Caspar.

Sørbø, J. I. (2004). Hans Skjervheim og pedagogikkens tredje [Hans Skjervheim and the pedagogical third]. In H. Thuen & S. Vaage (Eds.), *Pedagogiske profiler* [Pedagogical profiles] (pp. 303–319). Oslo: Abstrakt forlag.

Stramer-Brandt, P. (2018). *Projektarbeit in der Kita: Mit Checklisten und Kopiervorlagen* [Project work in the daycare centre: with checklists and copy masters]. Freiburg i.Br.: Herder.

Waller, T., Sandseter, E. B. H., Wyver, S., Ärlemalm-Hagsér, E., & Maynard, T. (2010). The dynamics of early childhood spaces: opportunities for outdoor play? *European Early Childhood Education Research Journal, 18*(4), 437–443. https://doi.org/10.1080/1350293X.2010.525917

Waters, J., & Maynard, T. (2010). What's so interesting outside? A study of child-initiated interaction with teachers in the natural outdoor environment. *European Early Childhood Education Research Journal, 18*(4), 473–483. https://doi.org/10.1080/1350293X.2010.525939

Wittmann, E. C. (2016). Die Grundkonzeption des Mathe 2000-Frühförderprogramms [The fundamental concept of the Maths-2000-programme for early support]. In E. C. Wittmann (Ed.), *Kinder spielerisch fördern – mit echter Mathematik* [To support children in a playful way – with real mathematics] (pp. 22–45). Seelze: Klett Kallmeyer.

PART 2

Counting

3

"EVEN BIGGER THAN THIS WORLD"

Young children thinking about numbers through their mathematical graphics

Elizabeth Carruthers, Alf Coles, and Janet Rose

Vignette

This vignette is a small part of the data drawn from a current PhD study on the pedagogy of children's mathematical graphics and comes from one of the portfolios of practice within the study. It was transcribed from a video-tape, which all the commentators have seen, and thus has the added dimension of the commentators being able to observe closely the children's space, place, and non-verbal reactions. The original study utilises qualitative research methods, drawing on ethnography (Hoey & Fricker, 2007). It is partly participatory (Manzo & Brightbill, 2007), requiring ongoing sensitivity to emerging ethical issues.

The vignette is centred in a state nursery school[1] in England. This nursery school is also a National Teaching School.[2] It is in an area of deprivation, one of the poorest wards in the inner city. The children are of a multifaith and multi-ethnic background, with over 17 different languages spoken within the families at the nursery school. This nursery school believes, very strongly, in children's autonomy and self-initiated learning. At the beginning of the morning, children come into the nursery and go to their key nest group, which is a nurturing space belonging to the children and with the children's photos, graphics, and books of interest. The children and teacher engage in what the nursery calls "Talk Time" where the focus is on conversation. The expectation is that each child will be given the opportunity to engage in a conversation with their key person (Goldschmeid & Jackson, 1994) and other children. The key nest is a group of about twelve children, with one qualified early years practitioner/specialist or qualified teacher (key person).

The teacher has put out a range of maths equipment following on from children's previous enquiries about infinity. For example, there is a range of markers, clipboards, an abacus, metre rulers, and a number grid 1–100. There are two whiteboards at children's height: one is blank and the other the teacher has drawn a grid on, for the children to use as they want. On the floor, there is a large long piece of white card, where the teacher, on one part of the card, has written numbers from 100, in one hundreds, to 900. On another part of the card, he has written numbers from 100, adding a zero each time, to one million. These resources are seen as a stimulus for discussion and individual thinking.

There are five children sitting together with the teacher on the floor, including four 3-year-olds who have English as an additional language. The teacher says to Shelby (3 years and 11 months), "Ben has made a super number. Can you see that, Shelby? I think he's making a super number." Ben (4 years and 5 months) is on the floor writing beside the numbers on a white card. He is drawing a line of zeros on the card and says "banana." The teacher asks, "Why did you do a banana?" Ben laughs and says, "It's a banana number." The teacher replies

"It's a banana number? Wow! You know what, Ben? I wonder if that's because last week we counted how many people like bananas. Do you remember? Have we got a number that big on our number line here? ... 50 million quillion."

Ben then says "trillion." The teacher asks, "Are you starting from over here?" and points at the corner of the card, "all the way round?"

Aaron (3 years and 6 months) now arrives with his grandma and approaches the teacher with his lunchbox. The teacher says to Aaron, "I am just doing

FIGURE 3.1 Ben

FIGURE 3.2 Aaron

some counting, Aaron … look, what does it say, 150 million quillion?" Aaron does not reply. The teacher says, "Do you like numbers, Aaron? … Come and see what Ben's doing, 150 million quillion." Ben is now at the whiteboard, filling it up with zeroes. The teacher says to Ben, "I have never known that number before. How many zeros do you have in your number?" (Figure 3.1). Ben says, "a quillion is even bigger than this world." The teacher replies looking at Rashad (3 years and 9 months), "Bigger than this world! Did you know that, Rashad, a quillion is bigger than this world?" Rashad points to the whiteboard. The teacher says to Rashad, "There are zeros on there."

Aaron's grandmother now joins in and asks, "How many zeros are there, Aaron? How many zeros are there in a million?" Aaron does not reply or look at his grandmother, and she adds "Count and tell!" The teacher looks at Aaron and asks – referring to the white card, on the floor where the number

FIGURE 3.3 Ash

a million is written – "Do you see a million there?" Aaron's grandmother says, "He knows." The teacher asks Aaron's grandma if she knows what a million looks like and she replies, "Yes, million, billion, I'm a physics student."

Meanwhile, Aaron writes numbers in a grid on a whiteboard. The teacher now goes over to Aaron, sits beside him, and comments on his numbers: "Aaron's putting numbers in the boxes, aren't you Aaron? … You see that, Ash? Look what Aaron's doing. He's found some boxes and he is putting numbers in them." Aaron writes a number from one to six in each box, then repeats this, continues with the number seven and then adds another seven. He then repeats the sequence in the remaining spaces on the whiteboard (Figure 3.2). The teacher says, "That's interesting, Aaron. You did numbers, didn't you? And you did two sets of numbers. Can you see? You did." The teacher now points at each number and Aaron says the numbers, "One, two, three, four." Aaron pauses and the teacher says, "Two fours." Aaron continues to count, "Five, six, seven." The teacher says to Aaron, "You did another seven, didn't you? … And then started again." Finally, the teacher says, "See, you did two lots of seven, didn't you?"

Now, Ash (3 years and 11 months) starts to write on the whiteboard. She writes one to ten vertically down the board. The teacher says, "What number do you think she is going to go up to? Look at this, Rashad!" Ash stops, and says quietly to herself, "One, two, three, four, five, six, seven, eight." Then she continues to write nine and ten. The teacher says, "You did it all the way down to there, Rashad, before you have a go." Rashad is pushing to have a go and the teacher asks, "Shall we just ask Ash? Can you go past ten?" Ash rubs out the numbers, perhaps because she ran out of space, vertically. The teacher says,

"Oh, you're going to rub them all out, are you? She went to ten, didn't she? And she said she wanted to go to eleven. … Start with eleven, Ash, and see if you can go past eleven." Ash continues to 15. The teacher comments, "Ah, look, can you see that, Rashad? Ben, I am just looking at what Ash is doing. You might be interested in this. She is going all the way down." Ash continues with 16 to 20, watched by the others (Figure 3.3). The teacher comments, "That's brilliant, Ash. Look she's going to rub those out. She's not going to go more than 20, is she, Ben?" Ben shakes his head. The teacher says, "You don't think she is?" Ash says "twenty-one" as she writes it on the whiteboard.

Reflective questions

1   What mathematical concepts are the children individually working with?
2   In what ways does the teacher support the children's thinking and exploring?
3   Does this vignette show a way of teaching that is possible in your context?

Commentary 1: accessing a culture of mathematicians

Author: Alf Coles

In the last 20 years, in England, there have been many changes to the statutory guidance in mathematics for early years, and the current version is more prescriptive, with less emphasis on play than previous iterations (Whitbread & Bingham, 2011). This has brought tension within the early years sector, with concerns that schoolification now dominates the discourse (Whitbread & Bingham, 2011). The present curriculum document (DfE, 2017) places emphasis on number, space, and shape:

> Mathematics involves providing children with opportunities to develop and improve their skills in counting, understanding and using numbers, calculating simple addition and subtraction problems; and to describe shapes, spaces, and measure.
>
> *(p. 8)*

More recently, there has been a prominence of Asian-style mathematics, *Mastery Mathematics* (Drury, 2014), which the government has funded with the aim of tackling the historical tail of underachievement in mathematics in England. Although this approach to mathematics was originally targeted at school children from Year 1 onwards, it has filtered into early years, with mixed reviews (Boylan, 2019).

Counting

Given the context above, I would like to point to two features of the vignette in this commentary, one related to counting and one related to ritual. Bishop (1988b) identified counting as one of six "'environmental' activities" (p. 147), common to all cultures, from which mathematics derives, and counting is clearly the focus of the latter part of the vignette. The earlier part, with Ben, is perhaps closest to one other of Bishop's universal categories, "playing" (p. 149), albeit without any explicit rules.

In order to help think about the kinds of counting we can observe, I will use a distinction between forms of counting. We can count a number of "things," perhaps pointing as we go, *or* we can purely count, saying the number sequence. Gelman and Meck (1983), in their influential research, proposed that early experiences with number should centre around the counting of things. Nathalie Sinclair and I have argued (Coles & Sinclair, 2018) that there has been an overemphasis on such counting in the early years curriculum, in England and Canada and perhaps elsewhere, at the expense of simply counting for counting's sake, which can lead to activities such as noticing patterns, playing with word structures, and seeing how high you can get. Another way of saying this is that the cardinal aspects of number (relating to "how many" questions) have been overemphasised at the expense of also nurturing a more ordinal awareness of number (relating to "what comes next" questions).

The vignette begins with the teacher commenting on a number Ben has "made." There is a sense here of a number as a "thing" in itself, perhaps something that needs to be brought into existence through writing or saying. Ben gives his number the name "trillion" and later on comments, "a quillion is even bigger than this world," again invoking a sense of a number as an object in its own right. There is an absence of a referent for Ben's number, a fact associated with pure counting. In other words, there is no sense here of needing to find a quillion objects to match the number Ben has written. The numbers are being invoked and contemplated and played with for their own sake.

The action moves to Aaron, who is writing numbers in boxes (1 to 7). The focus for Aaron seems to be on the number sequence, again a form of pure (rather than transitive) counting. Ash joins in and writes the numbers 1 to 10. Encouraged by the teacher, "Can you go past ten?" Ash continues to 20 and then 21. As with Aaron, the focus here is not on linking the numbers to objects; the numbers' cardinality is not being spoken about. Instead, the focus is on the sequence itself. Ordinality is in play, in a kind of pure counting.

What I see in the vignette is consistent with the kind of practice that, in Coles and Sinclair (2018), we were calling for where children's enjoyment with numbers, *as they relate to other numbers*, is encouraged and nurtured. There is a joy in exploring large numbers and I see no "dangers" (as has sometimes been expressed to me) in children playing with numbers that they cannot "know" in a cardinal sense. Bishop (1988a) suggested that mathematics has some mystery as a result of its abstractions taking one away from contexts, to the point where we "develop meanings within mathematics" (p. 187). In any staged conception of learning or development, arriving at abstractions divorced from context is the reserve of maturity. And yet, in this vignette, I would argue we observe three-year-old children doing the work of precisely this development of meaning within mathematics. The numbers they are working with have meaning only in relation to each other, from what we can observe. And, I would argue, this is how mathematicians work with numbers most of the time. In other words, I could see the vignette as a process of enculturation into the practices of mathematicians. There should be no surprise that children of this age are able to work with abstract ideas and relations since language itself is abstract (Gattegno, 1974). Following Gattegno (1974), I might even argue the children are making use of the power of engaging with mathematics as a language.

Ritual

The second lighting on the vignette I want to consider concerns the notion of ritual. An influential use of the idea of ritual, in mathematics education, comes from the work of Anna Sfard (2008) in which she distinguishes ritual engagements in mathematics from an explorative engagement. Explorations aim to change or get to know the world. The aim of rituals is social, "the whole point in the ritual action is that it is strictly defined and followed with accuracy and precision so that different people can perform it in identical ways (possibly together)" (p. 244). Sfard also adds that rituals are "about performing, not about knowing," because there is "no room for a substantiating narrative." On this view, rituals are a necessary starting point for engaging in mathematics, which mature into explorations, at which point children might come to understand what they have been doing previously in a ritualised manner.

Again, with colleague Nathalie Sinclair (Coles & Sinclair, 2019), I have proposed the notion of "ritualisation" from the work of Catherine Bell (1991) as an alternative idea to that of ritual, where for a ritualisation there is not the sense of activity occurring in an unthinking or purely performative manner. We suggested three components of ritualisation: the activity is set apart as distinct and privileged; the environment in which the activity takes place is symbolically structured; there is little or no attempt to bring what is being done across the threshold of discourse. Ritualisations are also about becoming part of a culture, and the symbolic structuring of the environment might be about religious iconography or sets of values. Bishop (1988a, p. 187) refers to the "symbolic conceptual structures" that make up mathematics, which suggests that ritualisation practices might be particularly appropriate in learning mathematics and entering the culture of mathematicians.

I want to argue that the activities around number in the vignette have qualities of ritualisation and will take each of the components in turn. There are a number of interventions by the teacher which suggest that the activities around numbers in which the children engage are both distinct and privileged. He says: "Ben has made a super number"; "Do you like numbers?"; "Come and see what Ben is doing"; "I have never known that number before"; "Aaron's putting numbers in the boxes"; "Look what Aaron's doing"; "That is interesting, Aaron"; "What number do you think she is going to go up to?" In these comments, the teacher is drawing others into activities around numbers, pointing out the activities taking place relating to numbers and, at times, positively reinforcing what the children are doing as interesting. There is a sense of privileging the number work taking place.

The second feature of ritualisation is that activity occurs in a symbolically structured environment. The children are grappling with the structure of number naming in English. An artefact present in the room, which supports awareness of this structure, is the number line. In Aaron's and Ash's actions, the number names are being invoked in a culturally appropriate order, although the teacher allows some divergence (e.g. Aaron writing two 4s). The environment affords the children some feedback in relation to this structure of number names (e.g. the teacher pointing out Aaron's two 4s).

Finally, in relation to ritualisation, there is the question of what is being brought across the threshold of discourse. The teacher is not seeking for the children to explain what they are doing. The teacher is not wanting anyone to say "why" 21 follows 20, nor, for example, what the "2" and the "1" represent in the number 21. The activities of the children are sufficient in themselves with "little or no attempt" to introduce a realm of discourse *about* them.

Conclusion

In conclusion, what I interpret in the vignette is a joyful, active, and playful engagement in numbers. The teacher is rarely evaluating (except for the occasional "brilliant"). Rather, he is anticipating the children's actions and verbalising these anticipations: "I think he's making a super number"; "What number do you think she is going to go up to?"; "She's not going to go more than twenty, is she, Ben?" There is something worth exploring, and the teacher seems attuned to catching that excitement. I read in these practices a trust that, if children are allowed to follow their own lines of enquiry, they will become ever more sophisticated in their use of number, given (and these are, for me, crucial caveats) the symbolic structuring of the environment, the sensitivity of the teacher, and the way in which numbers are being invoked as entities that have relations to each other (as well as objects). At a time when the curriculum in England is being conceptualised in terms of small steps of learning, it seems important to keep in mind that learning is not nearly so neat and children's interests are often tilted towards the complex, the abstract, and that there is much that we, as teachers, still need to learn about ways of giving space to such inclinations.[3]

Commentary 2: developing a community of mathematical enquiry

Author: Elizabeth Carruthers

I was the headteacher of the nursery school in this vignette. I am passionate about children's mathematics and led this English nursery school's developing practice, through teacher/practitioner research. The nursery school has documented many scenarios of the mathematical intentions of the children. Each one is different (see also Chapter 11). My response to the vignette is from a sociocultural stance (Rogoff, 1990; Vygotsky, 2012) where children are believed to learn from each other, their teacher, the school, their families, and their communities. At the heart of the mathematical learning in this vignette, there seem to be two key tools that children are drawing on. One is language through talk and the other, which is more dominant, their chosen written mathematical graphics (Carruthers & Worthington, 2005). I am mainly commenting on the pedagogical aspects of this vignette.

The beginning of the nursery day

Intentionally, at the start of the nursery day, a space and time for conversation and communication has been constructed and this is the first part of the pedagogical structure. Without this quiet space, with little interruptions, the children will find it difficult to be heard or listened to. That might be a simple and obvious statement; however, planning to have that time and space, within a busy play environment, is not an easy managerial task (Fisher, 2016). This is a regular time, at the beginning of the morning, and children are attuned to this. They and their families know that the teacher is available and listening. Within what we call *the key nest* (an area of the nursery that belongs to this group of children), the scene is set for enquiry, thinking, reflecting, and doing.

Attached teaching

The teacher knows the children and their mathematics, and has shaped the learning session around the children's previous enquiries about infinity. This knowledge of not just knowing the children, in general, but also knowing their individual mathematics is essential to the pedagogy in this vignette. The importance of the key person in early education in England is highlighted and supported by *Statutory Framework for the Early Years Foundation Stage* (DfE, 2017). It is grounded in Goldschmeid's work (Goldschmeid & Jackson, 1994), where she emphasises the importance of a special, caring relationship which is vital to a young child's well-being. This is what I refer to as "attached teaching" (Carruthers, 2017), and it is not only about care but tuning into the child's home and community knowledge (Moll et al., 1992). It is also a crucial way to understand young children's mathematical thinking.

At the time of this vignette, most children had arrived into the key nest area. The teacher had set up what might be termed *open suggestions through resources*. He is sitting on the floor so that he can be at the children's height for discussion. His positioning is not random, but a pedagogical strategy that aligns with the democratic practices of the nursery school. It is an aspect of attached teaching (Carruthers, 2015), being physically at the same level as the children. This may help the children hear and see the teacher as an equal, in the communication process and the search for meaning.

Developing a learning community

In this vignette, the teacher has not prepared a script to follow; there is no pre-planning, except for resources. It is about problems or investigations that arise in the momentum of present encounters (Lenz Taguchi, 2010). Infinity and larger numbers are a group enquiry, and the teacher encourages this collaboration throughout the vignette by providing a commentary. For example, "Aaron's putting numbers in the boxes, aren't you, Aaron?" He also invites the children to contribute: "Did you know that, Rashad, a quillion is bigger than this world?" and "What number do you think she is going to go up to?" He also acknowledges children's interests and addresses the children as thinkers: "I am just looking at what Ash is doing. You might be interested in this." This commentary also highlights and draws attention to each child's talk and graphics. This values what the children are doing, giving them confidence to investigate further. The teacher, through multiple ways of communication, seems to be establishing *a learning community* (Wenger, 1998) that highlights and encourages interactions between all learners.

The teacher is following the children's lead. They chose what they wanted to write and draw and, through this, led their thinking. He responded to the children's cues, as Goouch (2010) states, trying to unpick the complexities of the situation, within a dynamic intricate environment, and not only making a "contribution to thinking" but providing a thinking culture (Siraj, Kingston, & Melhuish, 2015, p. 7). The teacher has been influenced by Rogoff's (1990) work on guided participation and the importance of social interaction for learning.

Children's mathematical graphics

The mathematics in the vignette is not the government-set curriculum mathematics. For example, infinity is not on the standard curriculum for three- and four-year-olds in England. It is the children's mathematics (Carruthers & Worthington, 2006) and comes from their own experience, ideas, and connections they make (Carruthers, 2017).

At the beginning of the vignette, there is more communication through talk, but for the most part, children seem to be communicating through using their graphics or observing what is happening. An important point is that although there are other materials to use, such as an abacus, metre rulers, and string counting beads, the children mainly choose graphic materials to work with. Ben is writing zeros on the whiteboard, dancing in and out of talk with the teacher; he may be demonstrating his understanding of zeros and the connection with large numbers. He uses his own labels for the numbers, for example, "banana numbers," and his written zeros he says are a "million quillion" and he expresses his idea of infinity, "even bigger than this world." Aaron has been drawn to the empty grids on a whiteboard, and puts numbers in them, the written numbers he knows and is thinking about at this time. Ash seems to be challenging herself by writing as many known numbers as she can, in sequence. She talks quietly to herself as she does this, checking her count. She is very focused and is not distracted by the others, who are pushing to see what she is doing. She is communicating her written number knowledge to the teacher and her peers. It seems these children are free to express their thinking about the mathematics they are interested in. I feel there is a spirit of children's quiet mathematical engagement, and the place of children's mathematical graphics is not only a tool to talk about but it is conversation in itself.

Four of the children are still only three years of age, yet three of them use the writing and drawing materials with ease. They are becoming fluent graphic users. As Bishop (1988a, p. 187) discusses acculturation, "The induction of a person onto a culture which is in some sense alien" to their home lives, the children within the openness of the nursery are striding across the home culture and the less familiar nursery culture. This is an important bridge.

The nursery provides ongoing opportunities for free drawing, exploring symbols and signs, and experimenting with marks. The key nest space belongs to the children and, besides children's photos and books of interest, it is adorned with all kinds of nursery and home drawings, writings, and written communication. Therefore, the culture of written/drawing communication has been established, and I argue, this is the vital foundation vehicle that encourages the children to use graphic materials as communication.

Aaron and Ash do not talk to the teacher throughout the vignette, but they are intensively involved in writing numerals, in a state of focused concentration (Cysikzentmihalyi, 1990). Bruner (1983) states that when a child is thinking hard about what he is doing, he is not necessarily needing to talk about it; rather, he is doing it. Rose and Rogers (2012) discuss the importance of knowing the adult well and feeling that trust and confidence so that they are not afraid to share their ideas or, in the case of Ben, Ash and Aaron, write and draw their thoughts. This is what Ash, Aaron and Ben seem to be demonstrating as they put their mathematics on the whiteboards. They all used standard symbols to represent their thinking; however, they represented them in individual ways: Ben drawing a burst of zeros, Aaron concentrating on numerals 1 to 7 using the grid, and Ash deciding to use a vertical layout to write numbers down the board in a line to 21. Interestingly, both Rashad and Shelby were onlookers. They did not talk or use written communication. As Fisher (2016) explains, some young children choose to stand back and assess the situation first. It may be that Rashad was trying to understand and reflect on what was happening before he had the confidence to partake. Eventually, he appeared to become involved in what Ash was doing and eagerly waited to have his turn on the whiteboard.

The children's graphics in this vignette appear to be a spontaneous discourse; Ash and Aaron were not communicating their thinking through talk but by graphically exploring the use and position of numbers. Their graphics seem to have provided another way of communicating and making sense of mathematics, as Kress (1997) identified: young children need to have opportunities "to participate fully and productively in the making of their meanings, in the ways that make sense to them" (p. 151).

Sustained periods of reflective community enquiry

The powerful mathematical idea of infinity was a group enquiry, and the children had been continually thinking about and revisiting this concept for some weeks, with some wonderful reflections. The teacher would now facilitate the learning further by providing more whiteboard and grid opportunities for children like Rashad and Shelby, who were on the outside, to engage in their own representations of their mathematics. Siraj et al. (2019) found that one of the most useful pedagogical approaches was through "Sustained Shared Thinking." However, this enquiry about infinity was an ongoing, recurrent piece of group thinking, over an extended period of four months. This may be more useful for building mathematical concepts.

Commentary 3: acculturation and enculturation in mathematical understanding – children's and adults' "cultural capital"

Author: Janet Rose

This commentary reflects on two key elements from a cultural perspective, one that focuses on some aspects of the style of adult interaction with the child and the other that compares the vignette to the South African context, drawing in particular on the work of Feza (2016, 2018a, b). In doing so, it draws attention both to the universality of mathematical development across cultures and highlights the need for early years educators to take heed of the particular cultural capital young children bring to the nursery.

For me, the most remarkable feature of the vignette from a cultural perspective is how the children's activities and adult interactions could easily be transported into a South African preschool. However, caution must immediately be exercised given that such a translation assumes the transportation is into one of the more affluent, "Westernised" preschools that have evolved in this African country and which mimic British colonial traditions. It does not acknowledge a more indigenous depiction of mathematical development in a different cultural context.

South African children's enculturation and acculturation (the process of learning about their own, native culture and the process of adopting host cultural norms, values, and beliefs) of Western mathematical ideas must be viewed within the context of the country's colonial history, even within the post-apartheid era where the legacy of so-called Western education still firmly remains. In a report for the Human Sciences Research Council in South Africa, Feza (2018b) highlights the poor mathematical performance of South African school pupils as being rooted in the poor mathematical foundational knowledge of young children, of which the "Cinderella status" (i.e. low and somewhat neglected status) of early childhood education in South Africa is a contributing factor. Preschool provision in South Africa is variable, with the majority of children from lower income groups not accessing preschool education. Moreover, educators hold lower qualifications in poorer socio-economic areas and little attention is given to early mathematical development or the provision of a developmentally appropriate environment, which promotes mathematical understanding (Kuhne et al., 2012).

Given the relevance of language in supporting the development of mathematical understanding, a paper by Feza (2016) acknowledges the barriers that migrant and immigrant children face are equally evident in young South African children whose first language is not that of the preschool they attend. But interestingly, in one study, she has identified how five-year-old Xhosa children used only English numerals in their community, which facilitated their mathematical development in the English-speaking preschools they attended before school. This illustrates the extent of the acculturation process of Western language and mathematical practices into indigenous cultures in South Africa and the ease with which they are absorbed by young children. The same is perhaps evident in the vignette in the examples of Ben and Aaron, who appear to have a background which may be of a different culture and appear to be easily adopting Western practices that may differ from their home backgrounds.

On the other hand, elsewhere Feza (2018a) notes how young Xhosas' cultural games and the artefacts they use are infused with mathematical concepts and skills, and should not be neglected for the part they play in developing their mathematical understanding, which perhaps also draws attention to the acculturation process of indigenous mathematics into Western practices (or vice versa). She calls for a "culturally relevant pedagogy" that acknowledges such cultural capital (Feza, 2018a). Lenz Taguchi's (2010) ideas of learners being "of the world" in a "co-dependency" with the material and human world may be relevant here. But perhaps Feza's research mostly identifies that, whilst educators need to operate where feasible within the cultural capital young children bring to the nursery, it is nonetheless possible to identify the universality of mathematical principles, such as cardinality (as also evidenced in the vignette), inherent within different cultural contexts.

This supports Bishop's (1988a) argument that some aspects of mathematics might be deemed "culture free" in the sense that they are universal to all cultures. Moreover, the fact that many of the children, or at least their families, in the vignette are from different ethnic and invariably cultural backgrounds, gives credence to the universal or "international" element within mathematical development in young children across cultures. From my perspective, there does not appear to be any evidence of the potential tensions that might arise from the acculturation of children like Ash, whose background may be different from that experienced in the British nursery environment. Indeed, Ash and Aaron appear to have readily embraced Western counting structures and its symbolism.

At the same time, a second glaring caution needs to be made about assumptions regarding the cultural origin of the young children in the vignette and their enculturation process. It might be asked whether their focus on adopting so-called Western practices of mathematical graphics (such as the number symbols written by Ash) or their own playful representations (such as the "banana numbers") have emerged from their family cultural background or the acculturation into the practices of the nursery they attend. The same could be said of young South African children entering the nursery system in South Africa: what do they bring from their own indigenous culture and how much of this has been infused by years of colonialism and the Western-based apartheid and post-apartheid system imposed on the country's majority population?

Bishop's cautions are echoed by Worthington and van Oers (2016) who, like Feza (2018a), note how young children draw extensively on their personal cultural knowledge in pretend play, exploring and elaborating their mathematical knowledge within the context of their unstructured pretence and imagination. The importance of cultural context and mathematical understanding is also highlighted by Gifford (2014), who suggests that generalising results of children across countries is potentially problematic, and that the effects on the nature and rate of mathematical development may be context dependent on, for example, parental expectations and culture that values mathematical success. She cautions against making assumptions within international developmental comparisons. The South African context may be a case in point given the alarm that has been raised about young South African children's apparent underachievement in mathematics and its causes (Kuhne et al., 2012).

The challenges in drawing out the cultural connotations within the vignette can also be seen in the section related to Aaron's grandmother who joins the conversation about counting. In the video of the vignette, it is apparent from her clothing and accent that Aaron's grandmother may have originated from another country, but she has no hesitation in echoing and emphasising the Western emphasis on cardinal elements of mathematics. She heartily joins in the teacher's initial attempts to invite the children (amongst other things) to count the zeros, directly asking her grandson to "count and tell." In this respect, the grandmother is reflecting a didactic style of questioning that precludes the kind of open-ended, possibility thinking and dialogic teaching more commonly advocated by the literature (McMahon & Rose, 2019; Sylva et al., 2014). Siraj-Blatchford and Manni's (2008) article on adult questioning techniques revealed that over 90% of questions posed by adults were limited to closed questions, in this case requiring recall of a fact. The teachers' own style of questioning, at times, might be deemed to be "closed." However, it is evident from the vignette that the environment created by the adult facilitates the children to develop their own ideas, notice patterns, and make links in their understanding. So, whilst a recent review of the debate regarding the adult's role in young children's learning, particularly during playful activities, remains unresolved (Payler et al., 2017), nonetheless Rose and Rogers (2012) note the challenge of the adult role in creating a conceptual space and emotionally affirming context which enables the child to contribute his or her ideas, by listening carefully to what is being said and done.

There are frequent attempts made by the teacher to understand what the children are doing and to encourage their explorations and activities (note, e.g. his frequent use of "I wonder if …"). The teacher has largely set aside any "teacherly" desire or pressure to give or get the "right" answer. Instead, we see many examples of the teacher, for example, tuning into the child and showing genuine interest, clarifying their ideas and offering suggestions – all styles of interaction deemed to be more appropriate ways of engaging with young children's learning (Jones & Twani, 2014; Siraj-Blatchford et al., 2002), and that resonate with the work of traditional sociocultural theorists. Whether the advocation of such collaborative pedagogy might be deemed to be universal and culture free and equally relevant to, for example, the South African indigenous context, or is merely another Westernised imposition, remains in question.

Notes

1 Nursery schools in England are a unique early years' provision that are government funded. They have a tradition dating back to the 1920s, rooted in the work of Margaret Macmillan and Susan Isaacs (Edgington, 1998).
2 The title "National Teaching School" is awarded to a school that has been judged outstanding and has the capacity to be a centre of professional development, research, and initial teacher training: these schools are funded for their work.
3 Ros Sutherland's untimely death cut short her involvement in this writing, but she has been present for me in writing this commentary. Ros's passion was for social justice (Sutherland, 2007), and I know she would have enjoyed commenting on this vignette, perhaps seeing in it an opening up of access to the power of mathematical practices and culture.

References

Bell, C. (1991). *Ritual theory, ritual practice*. New York: Oxford University Press.

Bishop, A. J. (1988a). Mathematics education in its cultural contexts. *Educational Studies in Mathematics, 19*(2), 179–191. Retrieved from https://link.springer.com/article/10.1007/BF00751231

Bishop, A. J. (1988b). The interactions of mathematics education with culture. *Cultural Dynamics, 1*(145), 145–157. https://doi.org/10.1177/092137408800100202

Boylan, M. (2019). Remastering mathematics: mastery, remixes and mash. *Mathematics Teaching, 266*, 14–18.

Bruner, J. (1983). *Child's talk*. Oxford: Oxford University Press.

Carruthers, E. (2015). Listening to children's mathematics in school. In B. Perry, A. Gervasoni, & A. Macdonald. (Eds.), *Mathematics and transition to school: international perspectives*, (pp. 313–330). Sydney: Springer.

Carruthers, E. (2017). Children's mathematics. *Early Education Journal, 83* (Special Issue: Early Years Mathematics), 4–6.

Carruthers, E., & Worthington, M. (2005). Making sense of mathematical graphics: the development of understanding abstract symbolism. *European Early Childhood Education Research Journal, 13*(1), 57–79.

Carruthers, E., & Worthington, M. (2006). *Children's mathematics: making marks, making meaning*. London: SAGE.

Coles, A., & Sinclair, N. (2018). Re-Thinking "normal" development in the early learning of number. *Journal of Numerical Cognition, 4*(1), 136–158.

Coles, A., & Sinclair, N. (2019). Ritualization in early number work. *Educational Studies in Mathematics, 101*(2), 177–194.

Cysikzentmihalyi, M. (1990). *Flow: the psychology of optimal experience*. New York: Harper Row.

DfE (2017). *Statutory framework for the Early Years Foundation Stage*. London: Department for Education. Retrieved from www.gov.uk/government/uploads/system/uploads/attachment_data/file/596629/EYFS_STATUTORY_FRAMEWORK_2017.pdf

Drury, H. (2014). *Mastering mathematics.* Oxford: Oxford University Press.

Edgington, M. (1998). *The nursery teacher in action.* London: Paul Chapman.

Feza, N. (2016). Basic numeracy abilities of Xhosa reception year students. *South Africa: Language policy, Issues in Educational Research, 26*(4), 576–591.

Feza, N. (2018a). Black students' rich mathematical experiences: mathematics concepts and Xhosa cultural games for reception class. In D. Farland-Smith (Ed.), *Early Childhood Education.* Retrieved from www.intechopen.com/books/early-childhood-education/black-students-rich-mathematical-experiences-mathematics-concepts-and-xhosa-cultural-games-for-recep. http://dx.doi.org/10.5772/intechopen.81039.

Feza, N. (2018b). Poor mathematical performance of South African students point towards poor mathematics foundation of young children. *Report for the Human Sciences Research Council, Education and Skills Development.* Retrieved from www.mah.se/upload/FAKULTETER/LS/LS-seminarier/POEM/Feza%20POEM2.pdf

Fisher, J. (2016). *Interacting or interfering? Improving interactions in the early years.* Maidenhead: Open University Press.

Gattegno, C. (1974). *The common sense of teaching mathematics.* New York: Educational Solutions.

Gelman, R., & Meck, E. (1983). Preschoolers' counting: principles before skill. *Cognition, 13*(3), 343–359.

Gifford, S. (2014). A good foundation for number learning for five-year-olds? An evaluation of the English Early Learning "Numbers" Goal in the light of research. *Research in Mathematics Education, 16*(3), 219–233.

Goldschmeid, E., & Jackson, S. (1994). *People under three: young children and day care.* London: Routledge.

Goouch, K. (2010). *Towards excellence in early years education: exploring narratives of experience.* Abingdon: Routledge.

Hoey, B., & Fricker, T. (2007). From sweet potatoes to god almighty. *American Ethnologist, 34*(3), 540–599.

Jones, M., & Twani, J. (2014). Having real conversations: engaging children in talk to extend their language and learning. In J. Payler, J. Georgeson, & J. Moyles (Eds.), *Early years foundations: critical issues* (pp. 67–77). Maidenhead: McGraw Hill Education.

Kress, G. (1997). *Before writing: rethinking the paths to literacy.* London: Routledge.

Kuhne, C., O'Caroll, S., Comrie, B., & Hickman, R. (2012). *Much more than counting: supporting mathematics development between birth and five years.* Cape Town: The Schools Development Unit (UCT) and Wordworks.

Lenz Taguchi, H. (2010). *Going beyond the theory/practice divide in early childhood education.* London: Routledge.

Manzo L., & Brightbill, N., (2007). Towards a participatory ethics. In S. Kindon, R. Pain, & M. Kesey (Eds.), *Connecting people, participation and place: participatory action research approaches and methods* (pp. 33–40). London: Routledge.

McMahon, K., & Rose, J. (2019) Talk, narrative and sustained shared thinking. In D. Davies, A. Howe, C. Collier, R. Digby, S. Earle, & K. McMahon (Eds.), *Teaching science and technology in the early years (3–7)* (pp. 36–56). Abingdon: Routledge.

Moll, L., Amanti C., Neff, D., & Gonzales, N. (1992). Funds of knowledge for teaching. *Theory into Practice, 31*(2): 132–141. https://doi.org/10.1080/00405849209543534

Payler, J., Wood, E., Georgeson, J., Davis, G., Jarvis, P., Rose, J., Gilbert, L., Hood, P., Mitchell, H., & Chesworth, L. (2017). *BERA-TACTYC Early Childhood Research Review 2003–2017.* London: BERA.

Rogoff B. (1990). *Apprenticeship in thinking: cognitive development in social Contexts.* Oxford: Oxford University Press.

Rose, J., & Rogers, S. (2012). *The role of the adult in early years settings*. Maidenhead: Open University.

Sfard, A. (2008). *Thinking as communicating: human development, the growth of discourses, and mathematizing*. Cambridge, UK: Cambridge University Press.

Siraj, I., Kingston, D., & Melhuish, E. (2015). *Assessing quality: Sustained Shared Thinking and Emotional Well-being (SSTEW) rating scale*. London: Trentham Books.

Siraj, I., Taggart, B., Sammons, P., Melhuish, E., Sylva, K., & Shepherd, D-L., (2019). *Teachers in effective primary schools: research into pedagogy and children's learning*. London: UCL-IOE Press & Trentham Books.

Siraj-Blatchford, I., & Manni, L. (2008). "Would you like to tidy up now?" An analysis of adult questioning in the English Foundation Stage. *Early Years, 28*(1), 5–22.

Siraj-Blatchford, I., Sylva, K., Muttock, S., Gilden, R., & Bell, D. (2002). *Researching Effective Pedagogy in the Early Years (REPEY)*. DfES Research Report 365. HMSO London: Queen's Printer.

Sutherland, R. (2007). *Teaching for learning mathematics*. Maidenhead: Open University Press.

Sylva, K., Melhuish, E., Sammons, P., Siraj, I., & Taggart, B. (2014). *The Effective Pre-school, Primary and Secondary Education Project* (EPPSE 3–16+), DfE RR 354. Retrieved from www.gov.uk/government/publications/influences-on-students-development-at-age-16 or www.ioe.ac.uk/eppse.

Vygotsky L. (2012) *Thought and language* (E. Honfiman, G. Vaker, & A. Kozullin, Eds. and Trans.). Cambridge, MA: MIT Press.

Wenger, E. (1998). *Communities of practice: learning, meaning, and identity*. Cambridge, UK: Cambridge University Press

Whitebread D., & Bingham, S. (2011). *School readiness: a critical review of perspectives and evidence*. TACTYC Occasional Paper No. 2.

Worthington, M., & van Oers, B. (2016) Pretend play and the cultural foundations of mathematics. *European Early Childhood Education Research Journal, 24*(1), 51–66.

4

"LET'S ROLL THE DICE"

Exploring amounts, counting, transcoding, and the investigation of invariance and variance

Aljoscha Jakob Jegodtka, Lilith Schechner, and Kelly Johnston

Vignette

The day-care centre is located in a town of around 40 000 inhabitants in the state of Brandenburg in Germany. It offers places for 130 children, split into 14 early-childhood care (ages up to three years), 45 preschool (ages three years to school-age) and 71 after-school care places (school-children ages six to ten years).

The situation is an excerpt of a morning circle of a group of six children accompanied by one professional. The group consists of two boys and four girls, aged three to four years. The children were brought up in their native tongue, German. The socio-economic status of their parents is unknown; as is whether there are special education needs among the children. The educator underwent vocational training with state accreditation and is a young professional.

The morning circle, which lasts 22 minutes, begins with a song, during which the educator and four children hold hands and stand in a circle while two children sit on chairs alongside and watch. After the song is finished, the educator and all six children discuss and agree on the current weekday. Following a "good-morning-song," the educator and the children sit on the floor exploring mathematical concepts together:

The educator asks one child to fetch a red dice, which is around 10 cm³ in size, and then she asks the children what they notice about the look of the dice. The first aspect brought up by a child is its colour, "Red and black"; this is picked up by the professional and used to draw attention to its pips, "What is black on the dice?" Then they focuses on the number of pips on each face, "The pips. Each face shows different pips. How many are here?" Simultaneously, she places the dice in the centre of the circle. They all lean over the dice and start counting separately: One child touches each pip, while the others yell out the numerals, for example, "One, two, three, four, five!" The exploration of the number of pips on each face takes two minutes at most.

Then the educator shifts the focus to another mathematical aspect, "Everyone's allowed to throw the dice once. And the number of pips you see, that's how many bricks you fetch here. Wooden building bricks." Here, the educator points to a box containing wooden building bricks of various sizes that are on the other side of the room. The children start throwing the dice in turn, count the number of pips, and name the corresponding number, "Four. I get to fetch four." Then they run across the room to the box, count the corresponding number of wooden bricks, carry them back to the circle, and place the bricks in front of themselves. Transcoding from the dice roll into count and number word and from the number word into a number of wooden bricks, takes about two minutes.

As all involved children and the educator have thrown the dice and fetched the bricks, the topic evolves from counting to a comparison of the sizes of the bricks in each pile. The educator introduces this, "Ok, now we all take a look at our bricks. Everyone, grab yours." Instantly, a child responds, "The small one and a really big one." This idea is picked up by the educator, "I have gotten a really big one." The educator places her wooden brick (about 20 cm in length) in front of her, and a child places her four bricks (each about 5 cm in length) next to it. The educator responds by asking, "Oh, what does this look like?" This prompts a response by a child, "Same length," which is acknowledged by the professional, "That's the same length, exactly, excellent." The teacher continues by asking, "What happens if you take one (brick) away?" The child removes one wooden brick, leaving three smaller bricks next to the teacher's long brick and exclaims, "That's three metres long." The teacher asks, "three metres long?" This is met with affirmative sounds by one child. Another child takes away another brick and utters "two metres long." The educator directs the children's attention back to the comparison of bricks, "Which one is longer?" First, a child responds "two metres" and points to the two shorter wooden bricks, but then points at the larger brick, "That one." The exploration of the different sizes of wooden bricks continues for some time before returning to the number of bricks. "Do I have four?" asks one child. This is answered by the educator, "Exactly, you have four. And how many does Sofie have?" prompting an answer by one child, "two," and an objection of another, "I only had two." To conclude the discussion, the educator points out, "Even though I had a long brick, I only had one. Just like Mia."

In a final sequence lasting a couple of minutes, the educator and the children discuss the number of their bricks and raise questions of equality. Emily begins, "I only have two. Should I give you some?" to which the educator responds, "You can give Emma as many of your bricks as needed, so she has the same number as you. Let's see if you can do that." Another child – Sofie – takes up this suggestion, "We both have three. And now we both need …: One and one, then we both have four," she states, showing two fingers. Sofie runs to the box of bricks, fetches two more bricks, and places one in front of her three and one in front of the other child, who also had three. The children continue to establish equality for some time, replacing wooden bricks or fetching additional bricks, to yield equal numbers of differently sized wooden bricks.

Reflective questions

1   How can we promote experiential learning in early childhood with everyday objects such as bricks, dice, and pencils?
2   Do we take into account children's interests when designing activities for the classroom?
3   Is there anything that surprised you in the vignette? How would you have dealt with this?

Commentary 1

Author: Aljoscha Jegodtka

The German context: mathematics in the early years

In Germany, there is a *Common Framework of the Federal States for Early Education in ECEC* (Standing Conference of the Ministers of Education and Cultural Affairs of the Länder in the Federal Republic of Germany (KMK), 2004), which defines the basics of early childhood education and care. Here, a total of six educational areas are described, with mathematics falling within the educational area "Mathematics, Natural Sciences, (Information-) Technology." In this common framework it is stated,

> Children of this age have a great interest in scientifically representable phenomena of living and inanimate nature and in experimentation and observation. Therefore, children's childlike curiosity and natural urge to explore should be harnessed to acquire the developmental handling of numbers, quantities and geometrical shapes, mathematical precursor knowledge and skills. Closely related to this is the imparting of knowledge about the uses and functionalities of technical and information technology devices that characterize the daily lives of children, and about skills in the practical handling of them.
>
> *(KMK, 2004, p. 4)*

The pedagogy in ECEC institutions in Germany is social-pedagogically oriented. A holistic approach is always implemented in the German kindergarten.

The pedagogue sets out to address the whole child, the child with body, mind, emotions, creativity, history and social identity. This is not the child only of emotions – the psycho-therapeutical approach; nor only of the body – the medical or health approach; nor only of the mind – the traditional teaching approach. For the pedagogue, working with the whole child, *learning, care* and, more generally, *upbringing* (the elements of the original German concept of pedagogy: *Bildung, Betreuung and Erziehung*) are closely related – indeed inseparable activities at the level of daily work. These are not separate fields needing to be joined up, but inter-connected parts of the child's life.

(OECD, 2004, p. 19)

While the concepts of *Bildung* (education), *Betreuung* (care), and *Erziehung* (upbringing) are intertwined, there is a tension between children's self-education and a co-constructivist approaches. This tension is also reflected in the education plans of the German federal states. These are ultimately responsible for formulating curricula that are mostly compulsory for kindergartens.

The educational situation described in the vignette took place in a city with about 40,000 inhabitants in the state of Brandenburg. The education plan in Brandenburg, entitled *Principles of Elementary Education – Daycare in Brandenburg from Birth to the End of Primary School* (Minister for Education, Youth and Sports of the State of Brandenburg, 2016) describes the basics of educational work in the field of mathematics as follows.

The development of competencies in mathematics begins with the interest of children. They want to understand the objective world and explore it by acting. The child acculturates mathematics by, for example, sorting or counting bricks. She or he starts to say the number words and then something crucial happens,

The child learns that the series of number names can be transferred to series of objects, and by pointing to a single item and saying the first name and repeating this process with the next number, it can give the number of objects.

(Minister for Education, Youth and Sports of the State of Brandenburg, 2016, p. 25)

It is assumed that the development of mathematical competencies is based on active and meaningful construction by an individual, related first to everyday and environmental experiences. In this process, it is the job of educators to support the children, arouse curiosity, and explore with them mathematical and scientific phenomena. The starting point and basis of the active child-like learning process are the topics of the children. The children are observed regularly by the educators, providing an indication of approaches for support and promotion.

Currently, various options are being explored in Germany to help guide and support the development of mathematical skills in children. One trend is the development of specific support courses. They are aimed at smaller groups, especially children with developmental delays in the field of mathematics, and have a course character (Gerlach, Fritz, & Leutner, 2013; Krajewski, Nieding, & Schneider, 2007). This is in contradiction to the usual pedagogical practice in ECEC institutions. Oriented on the principle of the everyday integration of early mathematical education, another scientific tendency analyses the potential of specific games and picture books with mathematical content (Bönig et al., 2017; Schuler, 2013; Vogt et al., 2019). It can be shown that the playful examination of mathematical topics is helpful for the development of mathematical competencies. It is particularly helpful when educators play games with mathematical content with children and accompany them through activating questions. For example, if the children play a dice game and want to get as many pieces as possible, there can be a question such as, "How many more pieces do you have than me?" This focuses in particular on process-related competencies (Schuler & Sturm, 2019). Added to this, there is the analysis of mathematical interactions between early childhood educators and children. The more mathematical activities that take place between children and early childhood professionals, the greater the development of the mathematical competencies of children (Jegodtka et al., accepted).

In the vignette, we see a high-quality addressing of mathematics in kindergarten

The early childhood educator takes on elements of the child's life: dice and building bricks. The number of pips on the dice is used to support an important step in the development of a number concept: the connection of the number word with the number of pips on the dice. Here, the one-to-one correspondence of numeral and pips on the dice is discussed. At the same time, their number is determined and a meaningful counting strategy is used. According to Gelman and Gallistel (1978), there are five counting principles. In this situation, two of these principles – one-to-one correspondence and the principle of stable order of the numbers up to six – were discussed.

Also included in the game is the cardinal number principle. Here the number of elements is focused on. "How many do I have?" is the corresponding question. This is repeatedly raised by the children, answered, and so casually discussed. This is one, if not the central, step in the development of an appropriate understanding of numbers.

As a game – the children are allowed to roll in turn and then pick the appropriate number of building bricks – the transcoding from number to number word to number is practised by the children. Thus, another counting principle is worked out in a playful way: the principle of abstraction. This principle holds that it does not matter what is counted. The counting principles always remain the same. In this playful way, central mathematical concepts are developed by the children and in cooperation with the early childhood educator. It starts with children's interest, is in the zones of proximal development (Vygotsky, 1978), and gives the children a mathematical sense.

In addition, the children deal with the difference between number and quantity. In doing so, they develop an understanding of invariance in Piaget's sense. Thereby, they can grasp that sets of building bricks can be longer than others – and still contain fewer elements. Here too they acquire a central mathematical concept.

In summary, it can be stated that in the playful examination of objects of the child's living environment, building bricks and dice, they are accompanied and supported by the educator in their learning. This illustrates the holistic approach of a pedagogy of early childhood in everyday life and playful practical implementation.

Commentary 2

Authors: Aljoscha Jegodtka and Lilith Schechner

The Programme for International Student Assessment (PISA) findings published in Germany at the end of 2001 sparked a great deal of discussion about the quality of the German education system. Starting as early as the 1990s, Book Eight of the Social Code (SGB VIII, n.d. – law on child and youth welfare) stipulated that kindergartens, in addition to upbringing and care institutions, are to be regarded as (equally important) educational institutions for children. This had an impact on employees in that sector. First, educators have the task of providing high-quality care for children. Second, they have to be involved in the holistic educational development of a child socio-pedagogically. Third, it is their task to facilitate each child's individual development according to his or her current level. The classical understanding of kindergartens as institutions of *Erziehung* (upbringing) is being overcome slowly. Understanding education as a professional task in kindergartens, on the other hand, is slowly gaining ground. The responsibility for the quality of education and care in kindergartens rests with the individual federal states, so there are no nationwide domain-specific pedagogical regulations.

The nationwide implementation of educational plans for kindergartens at the beginning of the 2000s[1] emphasised the importance of kindergartens as places of education, and the qualification and further education of educators is increasingly becoming a focus. Essential parts of the curricula are, among others, mathematical and scientific areas of expertise.

Due to an increasing emphasis on the educational mission of the kindergarten, new challenges have arisen for kindergartens and educators. The expectations of the school and also of the parents for future schoolchildren put pressure on educators to transfer as much knowledge as possible (writing names, reading numbers, recognising quantities) to the children before they start school. On the other hand, this can become a stressful experience for the children, not only because of the intensity of learning they are faced with but also because it leaves little room for individual interests and needs that should be considered.

When kindergartens bow to this pressure and see themselves as mere knowledge mediators, educators do not fulfil their mandate to encourage and holistically promote, foster, and facilitate each child's individual development. Especially in kindergartens, it should be taken into account that children discover new things with all their senses and thus memorise experiences and ultimately gain knowledge.

Children have to see, try, understand. In German kindergartens, there is no systematic transfer of knowledge based on defined learning goals, as is the case in schools. The pedagogy in kindergartens follows a different approach. Nevertheless, the connectivity of the development of mathematical competencies between kindergartens and primary schools is of particular relevance (Wittmann, Levin, & Bönig, 2016). However, this refers to mathematical guiding principles and process-related competencies, or, as Bishop (1991) expresses it, the six universal activities. There are great differences in how kindergarten and primary school address the six universal activities. Pedagogy and content design differ considerably and yet should refer to each other. This is a big challenge for educators. Fortunately, children have a lot of fun with mathematics. This can easily be observed in the everyday life of the kindergarten: They sort by sizes, shapes, and colours. They compare themselves with other children in regard to size and age. They build long lines of snakes, count during hide-and-seek, count things off, and so on. This can be a starting point for conscious engagement with mathematical topics.

Many everyday situations have a mathematical reference and yet – as the vignette illustrates beautifully – educational opportunities should be developed which introduce children deliberately to mathematics. The example given by the vignette shows that the educator is aware that basic mathematical experiences in kindergartens must be very tangible. Mathematics in kindergartens is not abstract. Dice can be examined and by doing so, its peculiarities are discovered. The pips can be counted and with the help of building bricks, the numbers behind the pips are shown in concrete terms. The educator encourages children to make guesses, compare, and change amounts. The result is a mathematical discourse in the morning circle, and the children can playfully grasp numbers and quantities. They learn while playing (Elkonin, 2010).

Mathematics in everyday life takes place in every kindergarten. It is important to make oneself aware of what mathematics involves and to discuss it with the children. Mathematics and language belong together. Mathematics without language does not work.

In kindergarten, there are countless opportunities to talk about mathematical topics:

> "When will I be picked up?"
> "How many children are we today?"
> "We have to clean up, everything in its place: Where does something belong?"
> "Who is bigger?"
> "Who is older?"
> "When is my birthday?"
> "What happens on my birthday?"

The shapes of the tables, windows, and doors can be sorted and classified and much more: anything can be a starting point for mathematical education when the child is used to talking about shapes, time, sizes, lengths, sequences, patterns, and structures.

In the beginning, mathematics and science are, above all, phenomena that children curiously want to question, discover, and understand. Children perceive our world in a different way and are curious on their own. Educational professionals need to maintain and support this curiosity. We should listen to children's questions, listen to their thoughts, and let them discover the answers themselves. We can pick up and help develop children's ideas. We can enable them to carry out experiments by themselves and thus understand the relationship between X and Y. Even in free play, we can support them didactically by giving impulses, for example, to build towers even higher, tools to measure and to weigh and thus compare.

Early mathematical education is embedded in the relationship between the educator and the child. It is therefore important for educators to communicate with the children, to offer them closeness and safety, and thus to provide the child with a retreat from which they can explore and discover their environment in kindergartens. Educators have to support and promote the child's basic need for autonomy by offering choices and different scopes. Everyday life and its design are more important than all sorts of educational programmes. Educators are companions in the acquisition of knowledge, who must give the children time and space to make their own experiences and hypotheses, and maybe even discard them. Maintaining and promoting children's curiosity and play instinct, giving them room for development, is *the* central challenge for all educators. It is essential for those working with children in kindergartens to develop concepts which take into account the current research on child development. This is also of great importance for the education of educators. Especially in the field of mathematical-scientific experiences, many educators associate their own school experience with their pedagogical educational mission. It is important to realise that encouraging the learning of mathematics in kindergartens gives everyone involved the chance to discover and internalise basic mathematical experiences with fun and games.

Commentary 3

Author: Kelly Johnston

The Australian context – mathematics and early learning

The *Early Years Learning Framework* (EYLF) is the first mandatory national early learning framework developed and implemented within Australia (Department of Education, Employment and Workplace Relations [DEEWR], 2009; Sumsion et al., 2009). It provides educators with a general direction and overarching guidelines for the preparation of appropriate programmes and learning experiences offered within early learning services. There is flexibility for educators to create curricula that are responsive to the unique attributes of their contexts, families, and communities. The term "curriculum" as used in the EYLF was adapted from *Te Whāriki*, the Aotearoa/New Zealand curriculum policy for early childhood education (DEEWR, 2009; Ministry of Education, 1996). This interpretation of curriculum is more open ended than in traditional conceptualisations and extends to include all experiences, events, and interactions, both planned and unplanned, that occur within an early learning setting (DEEWR, 2009).The EYLF recognises the importance of the early years for children's development, the importance of culturally relevant intentional teaching, and also the value of establishing reciprocal relationships with families and connections with the wider communities (DEEWR, 2009; Leggett & Ford, 2013).

The *United Nations Convention on the Rights of the Child* (the Convention) (United Nations, 1990) also underpins the EYLF (DEEWR, 2009), with a strong focus on children's social and cultural rights and responsibilities. Another significant influence from the Convention is children's right to play and sense of agency (DEEWR, 2009). Contemporary thinking in relation to early learning in Australia suggests that children learn best in play-based situations that reflect their home cultures and experiences (Barblett, 2010; Ebbeck & Waniganayake, 2016). This approach also features strongly in the EYLF (DEEWR, 2009), which defines play-based learning as "a context for learning through which children organise and make sense of their social worlds, as they engage actively with people, objects and representations" (p. 6). Support for play-based learning exists within the context of intentional teaching, where educators are encouraged to engage children in active discussions about their experiences and thinking processes, to extend further their understanding and learning (DEEWR, 2009; Siraj-Blatchford, 2009). Respecting and incorporating children's views and voices in this way aligns with sociocultural theory, one of the four main theoretical perspectives that underpin the EYLF and highlights the importance of scaffolding as well as the promotion of social interactions to support children's learning and development (DEEWR, 2009; Rogoff, 1990;Vygotsky, 1978).The key notions of "belonging," "being," and "becoming" as conceptualised within the EYLF reinforce the idea that children have strong connections to their families, communities, and cultures, and that it is within these contexts that they develop a sense of identity, personal interests, and knowledge of their world (DEEWR, 2009).

The EYLF does not have specific content areas, such as those common to school curricula. Instead it acknowledges the integrated and holistic nature of learning in the early years and poses a series of broad learning outcomes. The EYLF encompasses the idea that mathematics is best learnt through everyday routines and experiences. To this extent, however, key mathematical concepts and processes are included in the EYLF definition of what it means to be numerate:

> Spatial sense, structure and pattern, number, measurement, data, argumentation, connections and exploring the world mathematically are the powerful mathematical ideas that children need to numerate.
>
> *(DEEWR, 2009, p. 38)*

The EYLF assumes or implies that early childhood educators have knowledge and understanding of mathematical concepts and processes. However, many educators do not see themselves as strong mathematical thinkers and may have limited training or experience in teaching mathematics. This is significant as educator's skills and knowledge influence their confidence as well as what they intentionally teach and what is included in the curriculum (Björklund, 2012; Perry, 2009).

Relation to the vignette

The vignette is an example that could have very easily been observed in an Australian early learning setting. Children are active participants in the play-based experience, leading and shaping conversation and investigations. The educator is intentional in their mathematical focus and in how she guided children's thinking. This practice aligns with the EYLF, which suggests children should have opportunities to "contribute constructively to mathematical discussions and arguments" (DEEWR, 2009, p. 39). The educator's approach also corresponds with the EYLF definition of curriculum implementation, where content is planned but also spontaneous and responsive to children's interests and prior knowledge. The combination of planned and spontaneous experiences is acknowledged as being of pivotal importance in effective mathematical teaching (Björklund, 2012; Clements & Sarama, 2018). Flexible and responsive pedagogical approaches challenge the long-standing issue of mathematics engagement being seen as "right or wrong" (Bishop, 1988). Here the feelings of "security and control" that mathematics can afford people (Bishop, 1988) still exists but in a more agentic, child-specific form. In the vignette, the educator promotes rich dialogue where ideas are invited, acknowledged, and extended. Children's exploration of number extended into length, showing an understanding of unit iteration and informal measurement (Bishop, 1988). This was presented in a way that was meaningful and relevant to the children, including items (bricks) with which they had prior experience and understanding.

Children were encouraged to use mathematical language and to explain their thinking relating to number, parts of a whole, and measurement (Bishop, 1988). Communication skills such as these are identified as being of key importance in supporting the development of mathematical thinking and learning (Björklund & Pramling Samuelson, 2013). Additionally, in this vignette, familiarity with resources enabled children to demonstrate what they intuitively understood and also enabled them to experience the joy of making unexpected mathematical discoveries (Bishop, 1988). A key focus of numeracy in the EYLF is to create positive attitudes towards mathematics, with the acknowledgement that these foundations are formed in the prior-to-school years (DEEWR, 2009). This aligns with wider research findings which indicate that children's positive dispositions and confidence in themselves as mathematical thinkers and learners have long implications for everyday numeracy and academic achievement throughout their lives (Clerkin & Gilligan, 2018).

Part of mathematics being meaningful and relevant to children means being appropriate for their level of development and also their ways of experiencing the world (DEEWR, 2009). Bishop (1988) refers to "enculturation," where attitudes towards mathematics in Western cultures are shaped from Piaget's ideas on the sensorimotor and pre-operational stages of development. He notes that it is at these early stages that different cultural approaches to mathematics emerge, which creates a false understanding of mathematics as universal. Clements, Fuson, and Sarama (2017) challenge the relevance of Piaget's stages, showing that children are capable of diverse and sophisticated mathematical thinking much earlier than had been believed. The example in the vignette shows an educator who sees children as capable and competent mathematical thinkers, which aligns with both the guidance and the overarching philosophy of the EYLF. Bishop notes that theory and ideology form the cultural basis of how mathematics is taught and experienced. In the vignette, like in Australian contexts, a number of key factors are considered and acted on to increase intentionality and relevance in teaching. These can include elements such as the environment, resources provided, language used, interactions and conversations, use of scaffolding, and adaptability and spontaneity in how learning experiences are allowed to unfold (Touhill, 2012). The educator showed confidence with integrating mathematical concepts and processes, which is linked to strong pedagogical practice and learning outcomes for children (Perry & McDonald, 2015).

Note

1 The GDR had curricula for kindergartens for a long time. This was different in the Federal Republic of Germany. After reunification in 1990, there were no kindergarten curricula in Germany for about 15 years. As late as 2004, the federal states started to introduce educational plans.

References

Barblett, L. (2010). Why play-based learning? *Every Child, 16*(3), 4–5. Retrieved from www.earlychildhoodaustralia.org.au/our-publications/every-child-magazine/every-child-index/every-child-vol-16-3-2010/play-based-learning-free-article/

Bishop, A. J. (1988). Mathematics education in its cultural context. *Educational Studies in Mathematics, 19*(2), 179–191. Retrieved from www.jstor.org/stable/3482573

Bishop, A. J. (1991): *Mathematical enculturation: a cultural perspective on mathematics education.* Dordrecht, Boston, London: Kluwer Academics Publishers.

Björklund, C. (2012). What counts when working with mathematics in a toddler-group. *Early Years, 32*(2), 215–228.

Björklund, C., & Pramling Samuelson, I. (2013). Challenge of teaching mathematics within the frame of a story – a case study. *Early Child Development and Care, 183*(9), 1339–1354. https://doi.org/10.1080/03004430.2012.728593

Bönig, D., Hering, J., London, M., Nührenbörger, M., & Thöne, B. (2017). *Erzähl mal Mathe! Mathematiklernen im Kindergartenalltag und am Schulanfang* [Let's tell maths! Mathematics learning in kindergarten and at the beginning of primary school]. Seelze/Velber: Klett-Kallmeyer.

Clements, D., & Sarama, J. (2018). Myths of early math. *Educational Science, 8*(2), 71. https://doi.org/10.3390/educsci8020071

Clements, D., Fuson, K., & Sarama, J. (2017). What is developmentally appropriate teaching? *Teaching Children Mathematics, 24*(3), 178–189. Retrieved from www.nctm.org/Publications/

Teaching-Children-Mathematics/2017/Vol24/Issue3/What-Is-Developmentally-Appropriate-Teaching_/

Clerkin, A., & Gilligan, K. (2018). Preschool numeracy play as a predictor of children's attitudes towards math at age 10. *Journal of Early Childhood Research, 16*(3), 319–334. https://doi.org/10.1177/1476718X18762238

Department of Education, Employment and Workplace Relations [DEEWR]. (2009). *Belonging, being and becoming: The early years learning framework for Australia.* Canberra: Commonwealth of Australia.

Ebbeck, M., & Waniganayake, M. (2016). Perspectives on play in a changing world. In M. Ebbeck & M. Waniganayake (Eds.), *Play in early childhood education: learning in diverse contexts* (2nd ed.) (pp. 3–23). Melbourne: Oxford University Press.

Elkonin, D. B. (2010): *Die Psychologie des Spiels* [The psychology of play]. Berlin: Lehmanns Media.

Gelman, R., & Gallistel, C. R. (1978). *The child's understanding of number.* Cambridge, MA: Harvard University Press.

Gerlach, M., Fritz, A., & Leutner, D. (2013). *MARKO – T: Mathematik und Rechenkonzepte im Vorschul- und frühen Grundschulalter – Training* [Training of maths concepts for 5- to 8-year-old children]. Göttingen: Hogrefe.

Jegodtka, A., Hosoya, G., Szczesny, M., Jenßen, L. Schmude, C., & Eilerts, K. (accepted). *Professionalität in der Gestaltung von Interaktionen in Situationen mit mathematischem Inhalt in der Kindertagesstätte und die Entwicklung mathematischer Kompetenzen bei Kindern: Resultate einer explorativen triangulierten Videographie-Studie* [Professional design of interactions in situations with mathematical content in daycare centres and the development of children's mathematical competencies: results of an exploratory triangulated videographic study].

Krajewski, K., Nieding, G., & Schneider, W. (2007). *Mengen, zählen, Zahlen: Die Welt der Mathematik verstehen (MZZ)* [Amounts, counting, numbers: to understand the world of mathematics]. Berlin: Cornelsen.

Leggett, N., & Ford, M. (2013). A fine balance: understanding the roles educators and children play as intentional teachers and intentional learners within the Early Years Learning Framework. *Australasian Journal of Early Childhood, 38*(4), 42–50. Retrieved from http://files.eric.ed.gov/fulltext/EJ854992.pdf12

Minister for Education, Youth and Sports of the State of Brandenburg. (2016). *Grundsätze elementarer Bildung – Kindertagesbetreuung in Brandenburg von der Geburt bis zum Ende der Grundschulzeit* [Principles of elementary education – daycare in Brandenburg from birth to the end of primary school]. Weimar: Das Netz.

Ministry of Education, (1996). *Te Whāriki: He Whāriki Mātauranga mō ngā Mokopuna o Aotearoa/Early Childhood Curriculum.* Wellington: Learning Media.

OECD (2004). OECD Country Note. *Early Childhood Education and Care Policy in The Federal Republic of Germany.* Retrieved from www.dji.de/fileadmin/user_upload/icec/reports/OECD_country_note.pdf.

Perry, B. (2009). Mathematical thinking of children in rural and regional Australia: implications for the future. In R. Hunter, B. Bicknell, & T. Burgess (Eds.), *Crossing divides: proceedings of the 32nd annual conference of the Mathematics Education Research Group of Australasia* (Vol. 1). Palmerston North, NZ: MERGA.

Perry, B., & MacDonald, A. (2015). Educators' expectations and aspirations around young children's mathematical knowledge. *Professional Development in Education, 41*(2), 366–381. https://doi.org/10.1080/19415257.2014.990578

Rogoff, B. (1990). *Apprenticeship in thinking: Cognitive development in social context.* New York: Oxford University Press.

Schuler, S. (2013). *Mathematische Bildung im Kindergarten in formal offenen Situationen – eine Untersuchung am Beispiel von Spielen zum Erwerb des Zahlbegriffs* [Mathematical education in kindergarten in formal open situations – a study on the example of games about number concept development]. Münster: Waxmann.

Schuler, S. & Sturm, N. (2019). Mathematische Aktivitäten von 5- bis 6-jährigen Kindern beim Spielen mathematischer Spiele – Lerngelegenheiten bei direkten und indirekten Formen der Unterstützung [Mathematical activities of 5 to 6 years old children when playing mathematical games – learning opportunities of direct and indirect support]. In: D. Weltzien, H. Wadepohl, C. Schmude, H. Wedekind, & A. Jegodtka (Eds.), *Forschung in der Frühpädagogik. XII: Interaktionen und Settings in der frühen MINT-Bildung* [Early childhood research. XII: Interactions and settings of early STEM education]. (pp. 59–86). Freiburg i.Br.: FEL-Verlag.

SGB VIII. (n.d.). Sozialgesetzbuch VIII: Kinder- und Jugendhilfegesetz [Social Code Volume VIII: Child and Youth Welfare Act]

Siraj-Blatchford, I. (2009). Conceptualising progression in the pedagogy of pay and sustained shared thinking in early childhood education: a Vygotskian perspective. *Education and Child Psychology, 26*(2), 77–89. Retrieved from http://eprints.ioe.ac.uk/6091/1/Siraj-Blatchford2009Conceptualising77.pdf

Standing Conference of the Ministers of Education and Cultural Affairs of the Länder in the Federal Republic of Germany (KMK) (2004). *Gemeinsamer Rahmen der Länder für die frühe Bildung in Kindertageseinrichtungen* [Common Framework of the Federal States for Early Education in ECEC]. Retrieved from www.kmk.org/fileadmin/Dateien/veroeffentlichungen_beschluesse/2004/2004_06_03-Fruehe-Bildung-Kindertageseinrichtungen.pdf.

Sumsion, J., Barnes, S., Cheeseman, S., Harrison, L., Kennedy, A., & Stonehouse, A. (2009). Insider perspectives on developing belonging, being & becoming: the early years learning framework for Australia. *Australasian Journal of Early Childhood, 34*(4), 4–13.

Touhill, L. (2012). Play-based approaches to literacy and numeracy. *National Quality Standard Professional Learning Program, 66*, 1–4. Retrieved from www.ecrh.edu.au/docs/default-source/resources/nqs-plp-e-newsletters/nqs-plp-e-newsletter-no-66-2013-play-based-approaches-to-literacy-and-numeracy.pdf?sfvrsn=4

United Nations (1990). *United Nations Conventions on the Rights of the Child.* Retrieved from www.ohchr.org/EN/ProfessionalInterest/Pages/CRC.aspx

Vogt, F., Hauser, B., Stebler, R., Rechsteiner, K., & Urech, C. (2019). Learning through play: pedagogy and learning outcomes in early childhood mathematics. *European Early Childhood Education Research Journal, 26*(4), 589–603. https://doi.org/10.1080/1350293X.2018.1487160.

Vygotsky, L. S. (1978). *Mind in society: the development of higher psychological processes.* Cambridge, MA: Harvard University Press.

Wittmann, G., Levin, A. & Bönig, D. (Eds.) (2016): *AnschlussM. Anschlussfähigkeit mathematikdidaktischer Überzeugungen und Praktiken von ErzieherInnen und GrundschullehrerInnen* [LinkM. Compatibility of mathematical pedagogical beliefs and practices of educators and primary school teachers]. Münster, New York: Waxmann.

PART 3
Measuring

5

HARRY'S LINE WORK

Bob Perry, Sue Dockett, Elspeth Harley, Sonya Joseph, Marie Hage, and Oliver Thiel

Vignette

Harry is a 4-year-old preschool boy in a rural town in Australia, less than 100 km from Adelaide, the capital of South Australia. He attends a preschool, which is both physically and administratively attached to an independent faith-based school catering for children from ages 3–12 years. Sonya is an experienced early childhood educator with a bachelor's degree in early childhood education who has worked at the preschool for a number of years and Marie is the longstanding Director of the preschool. The preschool uses a pedagogical framework, which incorporates facets of the Reggio Emilia approach (Edwards, Gandini, & Forman, 2012) and adheres to the Australian early childhood curriculum framework (Department of Education, Employment and Workplace Relations, 2009). The preschool program often utilises the outdoors, including a "bush block" [small natural wilderness area] which is easily accessible by children and educators. The approach to mathematics learning in the preschool is based on earlier work of the authors around powerful mathematical ideas (Perry & Dockett, 2008; Perry, Dockett, & Harley, 2012). In this chapter, the vignette was constructed by Sonya in consultation with Marie and is taken from the practice and documentation of the educator, with the literature mentioned above forming the theoretical and practical framework for its analysis.

The action behind the vignette commences when Harry seeks out Danette, the administrative officer for the preschool, who was at her desk. She is using a ruler to measure, and Harry takes some interest in this, asking Danette what she is doing and how she is using the instrument. Sonya's documentation of Harry's interactions with Danette and then with Sonya follows in the form of a learning story (Carr & Lee, 2012).

Harry enjoys going into Danette's office in the morning, they have a chat and Harry usually finds something interesting to explore. This morning

I [Sonya] ventured in to see what was happening. Harry was busy working with Danette's rulers! Harry knows about rulers because he has one at home with which he plays, but his is not like Danette's – his is wooden but hers is transparent and you can see through it. For example, you can see the numbers from either side of Danette's ruler. Harry knows the numbers and can see the patterns in them as he counts and measures. Danette explained to Harry how she uses the ruler in her work and gave him some advice on how to use it to get a steady, straight line. After some practice, Harry got busy making lines on paper with Danette's pencils.

Harry then talked to me about rulers and how you can use them to measure and to draw lines. I could see that Harry was highly motivated to continue his exploration of lines and so I offered him the metre ruler we have in storage. Wow! He was impressed and keen to get busy making some "BIG, BIG lines."

Harry also knows many stories, including "Jack and the Beanstalk." Harry recalls this story as he is measuring and drawing. Presently, Harry introduces the notion of measuring the beanstalk with a ruler and the challenges that might bring.

"It's Jack's beanstalk ladder," says Harry. "So high, high, whoa. Higher, higher, don't look down. If you look up, you won't fall off. Four pieces long now Sonya, maybe 100 steps." Harry and I count and there are 88 steps, a terrific estimate. Harry concludes, "I need more steps for 100. This is so the giant goes that way, not Jack's way. Jack's house is here and the giant's there."

FIGURE 5.1 Starting to draw lines

Harry begins with a single sheet of paper (Figure 5.1) and explains, "I made a ladder, a big one with the big ruler first. Then another piece so it gets higher and higher." Harry manipulated the tape and paper to continue adding sheets of paper for his work; piece by piece the tape was added. I noticed that Harry worked out that a longer piece of tape covered a bigger area and he didn't need to get up and down so often.

The beanstalk ladder grew to three large sheets of paper with a ladder running the full length. Harry then decided to add some colour before declaring the job complete (Figure 5.2).

I was very impressed with Harry's motivation and perseverance with this self-initiated task. As we worked together, opportunities to introduce and use mathematical concepts and terms were presented, including powerful

FIGURE 5.2 Three sheets of paper

mathematical ideas of number, measurement, argumentation, and spatial awareness. I could see the growth in Harry's fine motor control of simple tools like tape and pencils. He was able to think about a story he had heard before and transfer an aspect of that story to a visual representation. Harry interacted verbally with me, expressing his ideas and exploring how a pattern system (number counting) can be represented in different ways (including measurement where the rungs on the ladder formed the measuring units).

Thanks for sharing your work with me today, Harry.

Reflective questions

1 What was the role of the metre ruler in this play situation? How was it used and what did it inspire?
2 What different units of measurement were used by Harry throughout the play situation? How did the educator facilitate Harry's experience with different units of measurement?
3 How does the educator stimulate Harry's curiosity and perseverance in his play and exploration? What else could she have done?

Commentary 1

Authors: Bob Perry and Sue Dockett

Curricular and pedagogical contexts and frameworks used in early childhood mathematics education in South Australia

The curriculum and pedagogy in early childhood education in South Australia is framed by *Belonging, Being & Becoming: The Early Years Learning Framework for Australia* (EYLF) (Department of Education, Employment and Workplace Relations (DEEWR), 2009), Australia's first national early years learning framework.

> The Framework forms the foundation for ensuring that children in all early childhood education and care settings experience quality teaching and learning. It has a specific emphasis on play-based learning and recognises the importance of communication and language (including early literacy and numeracy) and social and emotional development.
>
> *(DEEWR, 2009, p. 5)*

The EYLF has three related elements: Principles, Practices and Learning Outcomes. There are five Learning Outcomes. Two of these, along with the key components relevant to mathematics, are:

Outcome 4: children are confident and involved learners

- Children develop dispositions for learning such as curiosity, cooperation, confidence, creativity, commitment, enthusiasm, persistence, imagination and reflexivity.
- Children develop a range of skills and processes such as problem solving, enquiry, experimentation, hypothesising, researching and investigating.
- Children transfer and adapt what they have learned from one context to another.

(DEEWR, 2009, p. 34)

Outcome 5: children are effective communicators

- Children express ideas and make meaning using a range of media.
- Children begin to understand how symbols and pattern systems work.

(DEEWR, 2009, p. 39)

There is only one extensive statement in the EYLF concerning mathematics and its learning.

> Numeracy is the capacity, confidence and disposition to use mathematics in daily life. Children bring new mathematical understandings through engaging with problem solving. It is essential that the mathematical ideas with which young children interact are relevant and meaningful in the context of their current lives. Educators require a rich mathematical vocabulary to accurately describe and explain children's mathematical ideas and to support numeracy development. Spatial sense, structure and pattern, number, measurement, data, argumentation, connections and exploring the world mathematically are the powerful mathematical ideas children need to become numerate.
>
> *(DEEWR, 2009, p. 38)*

Another important document related to mathematics in South Australian preschools is *Implementation Guidelines for Indicators of Preschool Numeracy and Literacy in Government Preschools* (Department of Education and Child Development (DECD), 2015).

> The indicators recognise that all children come to preschool with numeracy and literacy capabilities, which they demonstrate in their own unique way. They provide significant identifiers of children's numeracy and literacy learning and development at preschool. The indicators:
>
> - describe how a child sees, interacts with and explores their world
> - identify aspects of numeracy and literacy learning that can be observed in the day to day learning context of a preschool
> - are interconnected and not sequential.
>
> *(DECD, 2015, p. 5)*

The four numeracy indicators are:

- I explore and understand my place and space in the world
- I measure and compare my world
- I analyse, read and organise the data in my world
- I quantify my world.

(DECD, 2015, p. 9)

The approach to mathematics learning in Harry's preschool is based on earlier work of the authors around powerful mathematical ideas (Perry & Dockett, 2008). This is also the same source from which the EYLF drew the above-mentioned eight powerful mathematical ideas (DEEWR, 2009, p. 38).

Bishop's six fundamental mathematical activities

Another major impact on Australian mathematics education at all levels is Bishop's work on mathematical enculturation and, in particular, his six fundamental and universal mathematical activities (Bishop, 1988). "Harry's Line Work" relates directly to Bishop's *measuring* activity as well as to *designing* through the plans Harry made for the ladder. Harry's initial interest in Danette's ruler, his previous interaction with a ruler at home, and his ability to consider similarities and differences between the office and home rulers, including the numbers and patterns and construction material, shows that he is interested and wants to explore further. While he mainly used Danette's ruler to draw lines, he and Sonya did also talk about how rulers could be used to measure. Sonya's decision to introduce the metre ruler was perspicacious and led to further measurement language including "BIG, BIG lines." Harry continued to *play* (another of Bishop's activities) with the metre ruler until he made the link to his previous experience with "Jack and the Beanstalk." While details are not given in the vignette, he and Sonya talked about *measuring* the beanstalk and how that might be done. Harry's work around Jack's beanstalk ladder introduces many other of Bishop's activities. While *measuring* continues, there is *designing* (the ladder), *counting* (the steps), *locating* ("Don't look down," "Jack's house is here and the giant's there"), and *explaining* ("I need more steps for 100. This is so the giant goes that way, not Jack's way."). All of this action and thinking is going on within Harry's *play*, stimulated by Sonya utilising approaches derived from the preschool's application of the Reggio Emilia philosophy (Edwards, Gandini, & Forman, 2012), particularly provocation.

Links to curriculum

While all of Bishop's activities can be discerned in "Harry's Line Work," with a preponderance of examples linked to *measuring,* there are also links to the powerful mathematical ideas noted in the EYLF. While the "content ideas" of *Spatial sense, number,* and *measurement* can be linked directly to Bishop's *locating, counting*, and *measuring*, what are sometimes called the "process ideas" of *argumentation*, *connections*, and *exploring the world mathematically* (otherwise titled *mathematisation*) can also be observed. For example, Harry provides a sound argument about why one should not "look down": "If you look up you won't fall off." The introduction, by Harry, of the "Jack and the Beanstalk" story consolidates the initial play with the rulers and provides a change in direction for subsequent play and investigation. It is an example of the prevalence of *connections* between Harry's mathematical work and other aspects of his life. The entire episode reported in the vignette shows Harry *exploring his world mathematically*, both in his concrete and in his imaginary worlds. Links to the four "content ideas" can also be made through further analysis of the one activity reported in the vignette, but this is left as a challenge to the reader. Sonya's summary of Harry's learning in mathematics and a number of other important curriculum areas demonstrates the value of the activity which was derived from Harry's play as well as the value of an astute and knowledgeable educator who is willing to observe and provoke as appropriate.

Potential follow-up activities

Sonya has taken great care and provided a lot of detail about Harry's activity and learning in the vignette. She has offered an overall evaluation of Harry's learning but has omitted to tell either the reader or Harry what might happen next. This final step is often omitted from such documentation (Carr & Lee, 2012).

Harry has shown that he has sufficient number skills – there is some evidence that, with assistance, he can rote count to 100 and that he recognises numerals such as those on the ruler – for him to start investigating measurements in terms of the numbers on the metre ruler. His interest in the "bush block" adjacent to the preschool might mean that he is given a challenge to measure various natural features. Some of these might be on the ground – such as paths – while others might be standing up – perhaps smaller trees. His interest in using the ruler to draw straight lines could stimulate further design activities where he needs to bring together lines that he draws to make other shapes. An investigation of other stories that he knows from his "reading" may lead to new investigations where he can use other measuring instruments such as metronomes, thermometers, and clocks.

Sonya's pedagogical approach where she observes aspects of Harry's play before provoking him with a piece of equipment, a question, or a story can provide many opportunities for follow-up. Key to the success of these activities will be Harry's willingness to be actively involved and his "motivation and perseverance with this self-initiated task." Harry does not need to be told what to do, but will relish opportunities for creative play leading to opportunities where he can be provoked into further play and learning. In some cases, the provocateur might need to be an astute and knowledgeable educator. In other cases, it could be Harry himself.

Commentary 2

Authors: Elspeth Harley and Marie Hage

Introduction

The Early Childhood Centre (preschool) which Harry attended is located on the Junior Campus of an Independent Lutheran School situated in one of the original towns (founded in 1848) in a large renowned wine-growing region in South Australia. The region and town were established by German settlers and have strong cultural and community values. Australia's first bilingual newspaper, *The German-Australian Post*, was published in the region in 1848. Nearby to the town is believed to be the world's oldest continually producing commercial vineyard, although this might be disputed by some of the readers of this book.

The school and preschool reflect the strong cultural and community values of the region. The preschool is acknowledged as one of the highest-quality centres in South Australia, receiving an overall National Quality Standard rating of Exceeding (Australian Children's Education & Care Quality Authority, 2019). It offers an educational programme for children aged three to school age (around five years in South Australia). The curriculum is planned using:

- *Early Years Learning Framework for Australia* (Department of Education, Employment and Workplace Relations, 2009),
- *Lutheran Education Principles and Values* (Lutheran Education Australia, 2019),
- *Principles and Philosophies of Reggio Emilia* (Reggio Emilia Australia, 2018), and
- *Nature Play Philosophies* (Warden, 2015).

The vignette is taken from the practice and documentation of the educator, with the literature mentioned above forming the theoretical and practical framework for its construction and analysis.

A play-based approach to teaching and learning is employed at the preschool with opportunities for "intentional teaching" (DEEWR, 2009, p. 15) grasped as appropriate. Children spend regular and extensive time in nature (on a bush block near the preschool buildings) and much of their learning arises from their experiences in this setting. In mathematics, emphasis is given to learning powerful mathematical ideas (DEEWR, 2009; Perry & Dockett, 2008) with plenty of opportunities to explore, investigate, solve problems, and use mathematical skills and language. Much of the mathematics learning undertaken by children at the preschool comes directly from their play, provoked by peers and adults through interactions, conversations, imagination, and play.

Players in the vignette

The joint writer of the vignette, with Harry, is Sonya, an experienced early childhood educator who has worked in the preschool for several years. The other adult involved is Danette, the administrative officer for the preschool. Danette has no "official" role in the education and care of the children at the preschool but often interacts with them as the children are free to enter the office at most times during the day. Danette does not hold early childhood education qualifications but does live and breathe the preschool's philosophy and view of children as capable and articulate learners.

At the time of the events depicted in the vignette, Harry was about four and a half years old. While he had built friendships with several children at the preschool, he was often happier to play by himself or near to a familiar adult. Harry was always "busy" and active and had many imaginative and innovative ideas. When he began preschool at age three and a half years, Harry could not settle at activities. He was very interested in and curious about everything in the environment but very restless, often jumping from one thing to another. He had poor listening skills, which sometimes resulted in his appearing "aloof" in his interactions with peers. Harry presented as a very "eager to please" child and demonstrated a warm and caring nature with peers and educators. When asked about his restlessness, Harry's mother revealed to his early childhood educator that he had always been a "high energy" child and was a very poor sleeper. Occupational therapy helped Harry significantly, and his educator observed, "Harry is becoming stronger in his social and emotional well-being. He is beginning to know how and when to take the time to choose experiences that are calming and of a more personalised nature." This history of Harry during the year before the vignette in this chapter was written is included here to illustrate that Harry's activity, reflection, and insight demonstrated through the vignette had not always been evident but were becoming more regular. Hence, Sonya was willing to provoke Harry in his learning, assured that his developing independence and self-confidence would facilitate positive responses to increased challenges.

Prior to the experiences recorded in the vignette

Harry had experienced numbers in many different ways: counting, recognition of numerals, and use of numbers in his environment. He had used rulers at home both to "measure" objects and to draw lines. He had not shown that he understood aspects such as "centimetres" or the need to match baselines when measuring.

Harry was interested in woodworking and spent time at the bench hammering nails and cutting wood. He sometimes tried to make a named object – perhaps a plane or a boat – but usually he just joined wooden pieces together. Educators fostered his passion for woodwork and encouraged him with many experiences that not only supported mastery of his woodworking skills and creative thinking but also his mathematical understanding, including following a plan, measurement, matching shapes, sizes, and patterns.

Bishop's six fundamental activities

The key link between Harry's experiences reported in the vignette and Bishop's fundamental activities is *measuring,* although there clearly are aspects of *counting, locating, designing, playing*, and *explaining*. That is, all six activities are included to a greater or lesser extent. Even though the ruler is used, it is mainly utilised to draw lines rather than measure. However, some of Harry's use of comparative language – "So high, high, whoa. Higher, higher," for example, clearly fits into the measuring activity as defined by Bishop (1988).

Post the experience recorded in the vignette

The intervention of Sonya with the metre ruler really did get Harry started on his continuing adventure. He wanted to make "some 'BIG, BIG lines'" and he did. The link to "Jack and the Beanstalk" certainly stimulated continued investigation and imagination. The educator provided a metre ruler (which was exciting for Harry because he had never seen one before and it reinforced his notion of "big"), large sheets of paper, and tape (to tape the sheets together because he wanted it "super long"), as well as time and space and encouragement to bring his idea to fruition. Harry displayed his ability with the manipulation of number and the design of the ladder.

The learning story documentation, including photos and narrative, enabled Harry to share his learning with his family and peers and the broader community. The story could also encourage him to continue with some aspects of his learning that have been highlighted. For example, combining his interests in "Jack and the Beanstalk" and woodworking could lead to it being suggested that Harry not only draw the giant's ladder but that he try to make it from wood. His interest in the outdoors might lead to his designing a ladder for one of the trees on the bush block and perhaps even assisting in its construction. His practical use of estimation (of the ladder steps) could lead to many other estimation activities, especially in his preferred outdoor areas.

Summary

What began as a simple excursion to the "safe" space of Danette's office led Harry, with the assistance of an astute and observant early childhood educator, to many exciting mathematical adventures. He has been able not only to stretch his knowledge and skills but also to show what he can do in an open, play-based setting, grounded in his own experiences and imagination. The importance of Sonya's intervention at timely intervals cannot be underestimated. Young children will learn powerful mathematical ideas as long as they are encouraged to do so by knowing, flexible, and playful early childhood educators. Having access to an educator who has reflected on what she observed, analysed what might be possible within the preschool context, and stimulated the learner has benefited Harry immensely.

Commentary 3

Author: Oliver Thiel

The didactical sequence of teaching magnitudes in Germany

German mathematics education focuses on (abstract) objects like numbers, space, shapes, and magnitudes (Kultusministerkonferenz, 2004) rather than mathematical activities like counting, locating, designing, and measuring (Bishop, 1988). The concepts are related, of course. We use numbers for counting. Locating is to seek out and determine a location in space. We use shapes when we design something, and we measure magnitudes. Eudoxus of Cnidus (ca 390–337 BC) defined a *magnitude* as something that can be *measured* by using a smaller magnitude of the same kind as a unit (preserved by Euclid, ca 300 BC, book V).

Piaget's theory that the child's development follows mankind's evolution of knowledge throughout history (Piaget, 1928) had a strong influence on mathematics education in Germany. Teacher educators developed a *didactical sequence of teaching magnitudes* (Lauter, 1976). They claimed that teaching magnitudes should not start with using conventional measuring devices like rulers or measuring tapes. First, the children should make playful experiences with different magnitudes and develop the vocabulary that is needed to describe their experiences: something is long, wide, high, heavy, and so on. Next, the children should compare representations of magnitudes directly with one another. For example, they compare two sticks and see that one is longer than the other. Then, the children should compare objects indirectly by using arbitrary units like spans, cubits, paper clips, pencils, and so on, before they measure with standard units. That means, the children shall reinvent measuring before learning to apply conventional procedures (Radatz et al., 1998, p. 170).

The didactical sequence has been criticised (Peter-Koop, 2001). One issue is the order of tasks. Natural development starts with easy tasks, but research shows that it is much easier for children in primary school to use conventional measuring devices than to reinvent the measuring process. Many prefer using a ruler instead of unconventional tools (Boulton-Lewis, Wilss, & Mutch, 1996), and are more successful in using a ruler than other tools for length comparisons when a direct comparison is not possible (Nunes, Light, & Mason, 1993). The second issue is about children's understanding. Bragg and Outhred (2000) found in a longitudinal study (grades one to five) with 120 Australian children that most children in grade five were able to measure the length of a line correctly, but many of them still had problems with tasks that require a deeper understanding. Both "(a) counting informal units, unit marks, or unit spaces" and "(b) aligning the ruler and reading the scale" (Bragg & Outhred, 2000, p. 117) are strategies that do not indicate that students understand linear measurement.

FIGURE 5.3 Two bugs are sitting on two different branches and want to crawl to the raspberry. Which bug has the shortest way? (Eichler, 2004, p. 14; picture taken from Nakken & Thiel, 2014, p. 293)

Young children's understanding of magnitudes and measuring

As a solution, Peter-Koop (2001) suggests that teachers should not teach unconventional and conventional measurement sequentially but in parallel and repeatedly. They should take into account the children's prior knowledge and let them compare and reflect on the advantages and disadvantages of different measuring methods. This is certainly true for primary school, but there are consequences and applications for preschool, too. A major part of children's development of understanding magnitudes happens before children start compulsory school (Buys & Veltman, 2008; Peter-Koop & Grüßing, 2006). Eichler and Lafrentz (2004) studied 1,800 German primary school starters' prior knowledge about measuring (Lafrentz, 2004). In contrast to the older children's preference for standardised measuring instruments in the studies mentioned above, none of the six-year-olds used conventional measuring tools to compare lengths or areas. One task was to identify which one of two bugs had the shortest way to a raspberry (Figure 5.3). The branches were given as physical objects made of wire. Most children used body parts like span and thumb as informal units, and some judged by eye.

The results of this study suggest that we can describe the development of young children's understanding of magnitudes and measuring by five stages (Eichler & Lafrentz, 2004; Nakken & Thiel, 2014). The model is consistent with the learning trajectory for length measurement given by Clements and Sarama (2009). I use Harry's experiences from the vignette to explain the model.

1 *Rough comparison:* First, the child learns to estimate roughly whether one object is bigger than another. Harry judged by eye that the metre ruler is longer than the ruler he used before. Moreover, he worked out that a longer piece of tape covers a bigger area.

2 *Direct comparison:* Visual estimation works only when the difference in magnitude is big enough. In order to get more precise results, the child learns to compare two objects directly. Harry puts two sheets of paper side by side in order to check whether they have the same length.

3 *Indirect comparison:* For objects that cannot be moved side by side, the child learns how to use a movable third object to perform an indirect comparison. There was no need for indirect comparison during Harry's line work, but we can guess that Harry might have drawn the following conclusion: Harry's wooden ruler has about the same length as Danette's ruler. The metre ruler is longer than Danette's ruler. Thus, the metre ruler is longer than Harry's ruler, too.

4 *Measuring with arbitrary units:* The child learns how to use informal units to quantify magnitudes. Harry stated, "Four pieces long now Sonya, maybe 100 steps." Then, he and Sonya counted to find out that it had 88 steps. Piece and step are unconventional units. Harry learned that measuring involves counting units and he experienced that a smaller unit (step < piece) gives a larger value (88 > 4) for the same magnitude (the length of the ladder).

5 *Measuring with standard units:* There is a disadvantage with unconventional units. They may vary in size. A ladder drawn on four small pieces of paper might be shorter than a ladder drawn on three large pieces. One might argue that Harry's steps are not "units" since his rungs are not the same distance apart (Clements & Sarama, 2009). Therefore, people invented standard units like metre, litre, kilogram, and second. Harry knows the numbers and Danette explained to him how rulers can be used to measure. Even if Harry knows how to use a ruler as a conventional measuring device, he does not measure Jack's beanstalk ladder in centimetres, because there is no need to do that.

The Norwegian case

In 2017, the Norwegian government implemented a new curriculum for early childhood education. The old curriculum prescribed that preschools "ensure that children encounter different types of measures, units and measuring equipment in their play and everyday activities" (Ministry of Education and Research, 2006, pp. 27–28). Measures, units, and conventional measuring equipment are no longer mentioned in the Norwegian curriculum for early childhood education (Nakken & Thiel, 2014). Now, preschool "staff shall [...] create opportunities for mathematical experiences by enriching the children's play and day-to-day lives with mathematical ideas" (Ministry of Education and Research, 2017, p. 54) to "enable the children to gain experience of quantities in their surroundings and compare them" (Ministry of Education and Research, 2017, p. 53). With this in mind, educators will focus on helping children develop a sense of measurement rather than teaching them how to use measuring tools.

Possibilities for further experiences

How can we provide opportunities for Harry to discover and reinvent the measurement procedure and develop a sense of measurement? He is obviously interested in measuring and in "Jack and the Beanstalk." In addition to the woodwork suggested in the previous commentary, I suggest sowing beans. I suspect that Harry will be interested in observing how beanstalks are growing – even though they will eventually not grow as high as Jack's magic beanstalk.

- Harry can compare directly the heights of two beanstalks side by side.
- If Harry wants to keep track of the growth over time, he has to find a way to record the heights. How can he do this? Cutting a cord or drawing marks on a piece of paper could serve for indirect comparison. Wooden building blocks could serve as informal units. He will find his own way.
- Perhaps another preschool has beanstalks, too. How can the children compare the heights using only a phone for communication? On first glance, counting building blocks could work, but are we sure that the other preschool uses blocks of the same size? The children will discover a need for standard units.

Summary

Piaget (1972) points out that teaching something will keep children from inventing it by themselves, and Lorenz (1992) reminds us not to introduce standard units before the children discover the need for units that are universally available and understandable. Of course, even young children know measuring tools from their experience in daily life. When children ask, we can show them how these tools work, but we should not be surprised when they use them in unconventional ways. This will allow young children to make discoveries and to pave their own way by following their interests. The educators' important task is to provide appropriate opportunities that spark children's interest in exploring, investigating, and discovering.

References

Australian Children's Education & Care Quality Authority. (2019). *National Quality Standard*. Retrieved from www.acecqa.gov.au/nqf/national-quality-standard

Bishop, A. J. (1988). Mathematics education in its cultural context. *Educational Studies in Mathematics, 19*(2), 179–191. Retrieved from www.jstor.org/stable/3482573

Boulton-Lewis, G. M., Wilss, L. A., & Mutch, S. L. (1996). An analysis of young children's strategies and use of devices for length measurement. *The Journal of Mathematical Behavior, 15*(3), 329–347. https://doi.org/10.1016/S0732-3123(96)90009-7

Bragg, P., & Outhred, L. (2000). What is taught versus what is learnt: the case of linear measurement. In J. Bana & A. Chapman (Eds.), *Mathematics education beyond 2000: Proceedings of the Annual Meeting of the Mathematics Education Research Group of Australasia* (pp. 112–118). Sydney: MERGA.

Buys, K., & Veltman, A. (2008). Measurement in kindergarten 1 and 2. In M. van der Heuvel-Panhuizen & K. Buys (Eds.), *Young children learn measurement and geometry: A learning-teaching trajectory with intermediate attainment targets for the lower grades in primary school* (pp. 37–66). Rotterdam: Sense.

Carr, M., & Lee, W. (2012). *Learning stories: constructing learner identities in early education.* London: SAGE.

Clements, D. H., & Sarama, J. (2009). *Learning and teaching early math: the learning trajectories approach.* New York: Routledge.

Department of Education and Child Development. (2015). *Implementation guidelines for indicators of preschool numeracy and literacy in government preschools.* Adelaide: Department of Education and Child Development. www.education.sa.gov.au/sites/default/files/implementation-guidelines-indicators-preschool.pdf?v=1465356423

Department of Education, Employment and Workplace Relations (DEEWR). (2009). *Belong, being & becoming: the early years learning framework for Australia.* Canberra: Department of Education, Employment and Workplace Relations. Retrieved from https://docs.education.gov.au/documents/belonging-being-becoming-early-years-learning-framework-australia

Edwards, C., Gandini, E., & Forman, G. (2012). *The hundred languages of children: The Reggio Emilia experience in transformation* (3rd ed.). Santa Barbara, CA: Praeger.

Eichler, K.-P. (2004). Geometrische Vorerfahrungen von Schulanfängern [Primary school starters' previous knowledge about geometry]. *Praxis Grundschule, 2,* 12–20.

Eichler, K.-P., & Lafrentz, H. (2004). Vorerfahrungen von Schulanfängern zum Vergleichen und Messen von Längen und Flächen [Primary school starters' previous experiences of comparing and measuring lengths and areas]. *Grundschulunterricht, 51*(7–8), 42–47.

Euclid. (ca 300 BC). Στοιχεῖα [Elements]. D. E. Mourmouras (Ed.), Retrieved from www.physics.ntua.gr/~mourmouras/euclid/index.html

Kultusministerkonferenz. (2004). *Beschlüsse der Kultusministerkonferenz. Bildungsstandards im Fach Mathematik für den Primarbereich* [Resolutions of the Conference of (German) Ministers of Education. Educational standards in mathematics for primary education]. München, Neuwied: Wolters Kluwer.

Lafrentz, H. (2004). *Vorerfahrungen von Schulanfängern hinsichtlich der Konzepte des Vergleichens und Messens von Längen und Flächen* [Primary school starters' previous experiences regarding concepts of comparing and measuring lengths and areas]. Rostock: University Rostock.

Lauter, J. (1976). Die Behandlung von Größen im Unterricht der Grundschule [Teaching magnitudes in the primary school classroom]. In J. Lauter (Ed.), *Der Mathematikunterricht in der Grundschule* [Mathematics education in primary school] (pp. 85–102). Donauwörth: Auer.

Lorenz, J.-H. (1992). Größen und Maße in der Grundschule. [Magnitudes and measures in primary school]. *Grundschule, 11,* 12–14.

Lutheran Education Australia. (2019). *LEA vision, purpose and core values.* Retrieved from www.lutheran.edu.au/about/lea-vision-purpose-and-core-values/

Ministry of Education and Research (2006). *Framework plan for the content and tasks of kindergartens.* Oslo: Ministry of Education and Research. Retrieved from www.regjeringen.no/globalassets/upload/kd/vedlegg/barnehager/engelsk/frameworkplanforthecontentandtasksofkindergartens.pdf.

Ministry of Education and Research (2017). *Framework plan for kindergartens – content and tasks*. Oslo: Norwegian Directorate for Education and Training. Retrieved from www.udir.no/globalassets/filer/barnehage/rammeplan/framework-plan-for-kindergartens2-2017.pdf.

Nakken, A. H., & Thiel, O. (2014). *Matematikkens kjerne* [The core of mathematics]. Bergen: Fagbokforlaget.

Nunes, T., Light, P., & Mason, J. (1993). Tools for thought: the measurement of length and area. *Learning and Instruction, 3*(1), 39–54. https://doi.org/10.1016/S0959-4752(09)80004-2

Perry, B., & Dockett, S. (2008). Young children's access to powerful mathematical ideas. In L. D. English (Ed.), *Handbook of international research in mathematics education* (2nd ed.) (pp. 75–108). New York: Routledge.

Perry, B., Dockett, S., & Harley, E. (2012). The *Early Years Learning Framework for Australia* and the *Australian Curriculum – Mathematics*: linking educators' practice through pedagogical inquiry questions. In B. Atweh, M. Goos, R. Jorgensen, & D. Siemon (Eds.), *Engaging the Australian Curriculum Mathematics: Perspectives from the field* (pp. 153–174). Adelaide, South Australia: Mathematics Education Research Group of Australasia.

Peter-Koop, A. (2001). Authentische Zugänge zum Umgang mit Größen [Authentic access to quantities]. *Grundschulzeitschrift, 15*(141), 6–11.

Peter-Koop, A., & Grüßing, M. (2006). Mathematische Bilderbücher: Kooperation von Elternhaus, Kindergarten und Grundschule [Mathematical picture books: co-operation between parents, kindergarten and primary school]. In M. Grüßing & A. Peter-Koop (Eds.), *Die Entwicklung mathematischen Denkens in Kindergarten und Grundschule* [The development of mathematical thinking in kindergarten and primary school] (2nd ed.) (pp. 150–169). Offenburg: Mildenberger.

Piaget, J. (1928). *Judgment and reasoning in the child* (M. Warden, Trans.). London: Kegan Paul, Trench, Trubner.

Piaget, J. (1972). Some aspects of operations. In M. W. Piers (Ed.), *Play and development* (pp. 15–27). New York: W. W. Norton.

Radatz, H., Schipper, W., Dröge, R., & Ebeling, A. (1998). *Handbuch für den Mathematikunterricht 2. Schuljahr* [Handbook for mathematics education in second grade]. Hannover: Schroedel.

Reggio Emilia Australia. (2018). *Principles*. Retrieved from https://reggioaustralia.org.au/category/principles/

Warden, C. (2015). *Learning with nature: embedding outdoor practice*. London: SAGE.

6

"IT'S A LOT OF WORK"

A tailor's measuring tape in the doll's house

*Maria Figueiredo, Luís Menezes, Helena Gomes,
Ana Patrícia Martins, António Ribeiro, Myriam Marchese,
Isabel Soares, and Lynne Zhang*

Vignette

In a small public early childhood centre in Viseu, 25 children from ages 3–6 years attended the class of Isabel, an experienced early childhood teacher. The centre had two large rooms and is used by Isabel's group only. It is next to a primary school, but the children interacted mainly among themselves. Viseu is a medium, quiet city where most early childhood centres now have three to six groups but this one was still working with just one group. Two student teachers from the masters' degree in early childhood and primary education join the group from March until June. As part of their practitioner's research work, a tailor's measuring tape was left in the doll's house with no explanation or introduction. The ways children used it in their play were observed for three days and this vignette is based on the records made by the student teachers.

The activity room was big and well equipped with play areas – doll's house, grocery store, library, games area, construction area – and two big tables, mainly for teacher-led activities. The teacher did not follow a specific pedagogical grammar (Oliveira-Formosinho, 2007). The organisation of the learning environment and the pedagogy was based on the official guidelines (Ministério da Educação, 2016). Each play area was set up by the teacher at the beginning of the school year, so there was little children's participation in the set up-or significant change throughout the year. The activities were based on a daily routine, loosely decided, that included teacher-led moments and time for free play in the areas, usually during the afternoon.

A measuring tape commonly found in children's homes was placed in the doll's house where it could be easily seen as if it had been there before. No information was given to the children about it.

When the children started playing in the afternoon, the usual organisation and distribution were observed. The doll's house was a favourite area for the children. Many would go through it during the afternoon. Some of the children noticed the measuring tape and incorporated it in their play. This was done in different ways.

Some children play with the tape, but not with its social purpose of measuring. Rita, a four-year-old girl, says, "I'm going to hold my baby with this (the measuring tape) so he doesn't fall down." Francisca (girl, five years old) replies, "Don't, he has a belt for that." Rita adds, "Then I'll measure, sometimes mums measure."

Three boys, Guilherme and Tiago (both four years old) and Rui (five years old), play together. Guilherme pretends the tape is a rope and creates a scene with cowboys and thieves. He shouts, "I have to swing the rope to catch the thieves. I caught you!" He wraps it around Rui's leg. Tiago asks, "Guilherme, what are you doing?! That is for measuring!" Guilherme replies, "But I'm going to arrest Rui, I'm a cowboy." He starts pulling Rui to the floor while Rui cries, "Let me go … Help! Help me, Tiago!" The boys are laughing.

Another group of children, including Tiago, recognises the function of the tape, seeing it as a measuring instrument. Tiago puts the tape around himself and asks, "What is my size?" Vera, a five-year-old girl answers, "None. That's not how you do it." Tiago does not agree, "No. You see what is my size." Vera explains, "No, we won't see. It's a lot of work."

Some children seemed to associate measuring with the presentation of a number. This is the case with Rita, a three-year-old girl, who announces, "I am going to measure my baby. One, two, three, four, five … he measures twelve." She counted the numbers on the tape with her fingers, not putting the tape close to the doll to measure it.

Other groups of children play with the measuring tape, showing that they recognise this object as a tool for measuring length. Some associate it with the idea of a line. They also reveal knowledge about its use, associating "zero" with the beginning of the line and the number at the end of this line as the measure.

Barbara and Vania, two girls, four and five years old, have a conversation. Barbara asks, "What is this?" Vania answers, "A measuring tape, let me measure." Barbara is surprised, "It's to measure?" Vania explains, "Yes, look, I'll measure the table. You put the tape at the beginning of the table and then you just stretch it. And when it is where the furniture ends, you see the number that is there. Do you understand?"

Two other girls, Carla and Carlota, four and five years old, are measuring a board on the wall. Carla commands, "Hold it there." Carlota explains, "It's not like that, Carla. You have to put it on the zero and it has to be in a line." Carla wants to know, "And now how much does it measure?" Carlota says, "Measures a one, a four and a five."

Reflective questions

1   The first commentary suggests that "Putting the measuring tape in the doll's house [...] meant skipping steps in what the children needed to experience and understand before accessing a standard measurement tool." Is this an approach which should be encouraged or avoided in early childhood mathematics? Why?
2   In their play with the measuring tape, the children used many mathematical words, but were they "doing mathematics"? Explain.
3   What other everyday objects would you consider useful (to mathematics learning) to place in the children's doll's house?

Commentary 1

Authors: Maria Figueiredo, Luís Menezes, Helena Gomes, Ana Patrícia Martins, and António Ribeiro

Early childhood mathematics education in Portugal: perspectives from the curricular, pedagogical, and research landscape

Early childhood education (ECE) in Portugal focuses on the ages of three to six and is called preschool education. Before the age of three, there is mainly socio-educational provision, whereas preschool education is the first stage of the education system. The Ministry of Education is responsible for ensuring the pedagogical quality of teaching in all preschool education institutions (public and private). Since 1997, there have been Curricular Guidelines for Preschool Education but not a programme. The Guidelines were updated in 2016.

In both versions of the Guidelines, mathematics is included as a domain in "Expression and Communication," one of three content areas. The other two content areas are "Personal and Social Formation" and "Knowledge of the World." Besides mathematics, the content area "Expression and Communication" contains physical education, artistic education, oral language, and introduction to writing. The content areas serve as references for curriculum planning and evaluation (Ministério da Educação, 2016). Although there are three different areas, the emphasis is on articulation and the connections between them.

The teachers' intervention or pedagogy is enacted through the organisation of the learning environment and the planning of activities. The learning environment should foster rich play, which is highly valued in the 2016 version of the Guidelines. When organising the learning environment and planning educational activities, the teacher should create opportunities for learning in all content areas, including mathematics. The inclusion as a domain in the "Expression and Communication" area is based on the idea of mathematics as a language, crucial for the structuring of thinking. Its relevance in daily life and for future learning is mentioned. Both as a language and through its concepts and ideas, mathematics is seen as important for children to "make meaning of, know and represent the world" (Ministério da Educação, 2016, p. 6).

The relevance of early, informal learning in mathematics is acknowledged and valued. Therefore, methodologically, the Guidelines take as a starting point the interests, experiences, and daily life of the children. The teacher should find ways to observe and reflect on those, and then offer diverse and challenging experiences and help children reflect on and discuss them to support the construction of mathematical ideas. Problem-solving is valued and the learning environment is key to promoting challenging situations. The availability of resources for manipulation and representation is therefore considered very important in promoting opportunities for mathematical reasoning and communication (Ministério da Educação, 2016).

In terms of content and learning expectations, the 2016 Guidelines present four blocks for mathematics: numbers and operations; data organisation and analysis; geometry and measurement; and interest and curiosity in mathematics. Significant changes both in terms of content knowledge and didactical approach from the 1997 to the 2016 edition of the Guidelines have been signalled (Silva, 2019). In the *measurement* block, the focus moved to identifying measurable attributes in objects. This perspective was introduced in 2008 when a booklet about geometry and measurement (Mendes & Delgado, 2008), alongside one about numbers and data, was published by the Ministry of Education, to support teachers' curricular and pedagogical work. The trajectory and methodology suggested are very similar to, and inspired by, the TAL project (van den Heuvel-Panhuizen & Buys, 2005).

The Guidelines lay out a process that starts with identifying measurable attributes – both in play and through teacher-led activities – and then being able to choose a measurement unit (natural or standard) to compare objects to, translating that comparison into a number. This should begin with direct comparisons between objects and move to non-standard measurement units (pencils, steps, and so on) and, finally, to standard units (in activities that are meaningful for the children like cooking). The need to measure in real situations is highlighted to promote the understanding of the sense of measurement in daily life, as well as the need for standard measures. It is also suggested that those situations involve different physical quantities (length, weight, capacity, volume, time, and temperature).

The Guidelines for mathematics have been considered a thorough, scientifically sound and didactically up-to-date curricular document (Silva, 2019). Considering the argument that mathematical values are always present and shape teaching (Bishop, 1988, 2016), one absence is the clarification of values that support the content and methodological options. This is an important point since ECE in Portugal assumes a close relationship with families, and the transition of values from families or crèches to ECE centres needs to be addressed, particularly in mathematics. It is also relevant to consider the proximity and dependence between ECE mathematics and primary school mathematics. This is revealed in the organisation of content in the four mentioned blocks, as well as in several references in the text about future school trajectories. These suggest a close connection to the primary school curriculum, but also to the (non-explicit) values that inform it.

The research that is developed in Portugal about early childhood mathematics education (ECME) also reveals a strong relationship to research on further levels of the education system, particularly primary education (six to ten years old). For example, regarding measurement, the main references used in the few existing small-scale studies (Marchese, 2016; Monteiro, 2012; Moreira, 2018) are Boavida et al., (2008), Moreira and Oliveira (2003), and Ponte and Serrazina (2000), and a booklet about geometry and measurement (Mendes & Delgado, 2008). Boavida et al. (2008) and Ponte and Serrazina (2000) are important works directed at primary and basic education, without specifically contemplating ECE. This lack of research investment in ECME had already been pointed out by Rodrigues (2010). It can be concluded that at both the curricular and research level, in Portugal, ECME has close connections to school mathematics.

The tailor's measuring tape in the doll's house: two concurring lenses of analysis

In the case of measurement, the main sources available for teachers in Portugal suggest a strong emphasis on measurement by comparison as a means to develop an understanding about measurable attributes of objects, such as length or weight, and a sense of measurement as an activity with social relevance. Putting the measuring tape in the doll's house, from that perspective, meant skipping steps in what the children needed to experience and understand before accessing a standard measurement tool.

The different reactions and uses of the object revealed how for some children the tape could be used for purposes not connected to measurement, for example, cowboys. Most of the measurement-related episodes still showed children had incomplete knowledge of the measurement process, even when the object sparked measurement-related play actions, like pretending to measure or attempting to measure. We suggest these gaps between children's meaningful measuring activities, such as comparing and ordering, and the imitated measuring activities, with "adult" measuring instruments, should be gradually narrowed through the use of non-standard measurement units (like a step or a cup) (Buys & de Moor, 2008). This is also the path suggested in the Portuguese Curriculum Guidelines (Ministério da Educação, 2016).

On the other hand, the emphasis on play, with objects and sociodramatic interactions, explains the suggestion that standard measurement tools are introduced in ECE classrooms, connected to their daily life use so their relevance is known and understood by children (Ministério da Educação, 2016). It is a question of having socially relevant and authentic objects present and available for children to explore and play with (Moreira & Oliveira, 2003), not instructing on how to use them or expect that measurement is learnt through them. The opportunities to play with the measuring tape, besides revealing interesting knowledge about measuring, allowed for the exchange of experiences between children. This was true not only about the connection between that object and the idea of measuring but also in the last two episodes, on steps for measuring length with the tape.

The vignette has a clear connection to "measuring," one of six fundamental mathematical activities described by Bishop (1988, p. 183) as "quantifying qualities for the purposes of comparison and ordering, using objects or tokens as measuring devices with associated units or 'measure-words'." The children in the episodes were aware of measuring as:

- an activity ("measuring," "to measure"),
- part of their daily life ("sometimes mum's measure"),
- something they could perform, at least in a play context ("You see what is my size"; "I am going to measure my baby"; "let me measure"),
- something leading to a quantifiable result ("how much does it measure?"),
- associated with numbers ("One, two, three, four, five … he measures twelve"; "Measures a one, a four and a five"), and
- connected to the idea of size ("What is my size?").

The episodes show how the measurement device had meaning for the children, expressed in their recognition of it and the ideas about measurement that were listed. Bishop's (1988) list highlights the ideas of length, measuring instruments, and qualities. The activity of measuring itself was not so successful, as most of the children didn't know the details of using the tape ("it's a lot of work") and none of them could read the final result correctly ("Measures a one, a four, and a five").

The episodes seem to concur with the idea of measuring as a significant activity, with social relevance, that children feel familiar with and have experiences of. The episodes also support the relevance of a supported learning trajectory that focuses on the measurable attributes of the objects while at the same time making sure children feel confident with the idea of measurement and learning about it.

Possible steps forward

Measurement is a fascinating concept and activity. "'How much?' is a question asked and answered everywhere" (Bishop, 2016, p. 45). A further understanding of how children conceive measuring is hinted by the diversity of contexts mobilised in their play: mothers and their babies, body/clothes sizes, furniture and objects on the walls. In the more restricted indoor classroom, three different purposes and contexts for length measuring were present. A possible follow-up would be to introduce tools for other attributes, like a measuring jug, a timer, a scale, a thermometer, or even a ruler. Observing the play around them could contribute to discussions on the central position of the physical quantity "length" suggested by the TAL project (Buys & de Moor, 2008; van den Heuvel-Panhuizen & Buys, 2005).

In terms of practice, besides sharing and complementing ideas about where measuring is useful in daily life, starting with the examples presented by the children in their play, an important follow-up would be to clarify that measuring can be achieved through different ways. This would remove the attention from using the measuring tape to the act of measuring itself. It would also be a way of promoting multiple opportunities for measuring the same object with different units and/ or methods (Tyminski et al., 2008). In these opportunities, allowing children to make decisions about what they want to measure, how to do it, and with what are important considerations (Moreira & Oliveira, 2003). For any suggestion, the creation of an environment that supports children's exploration and confidence, as well as their sense of wonderment and perseverance, is crucial.

Commentary 2

Authors: Myriam Marchese, Isabel Soares, and Maria Figueiredo

Contextualisation

As part of the master's degree in early childhood and primary education, a small research project was developed. It looked into the ways children explored a measuring tape during their play. The observation of the moments in which children used the tape in their activities provided a window into their knowledge about measurement. The process started with a discussion about how literacy-enriched play environments suggest that authentic, complex objects of literacy should be incorporated into children's play (Christie, 2005). Observing children "write" the shopping list before going to the grocery store raised that discussion. The Movimento da Escola Moderna, a pedagogical model in Portugal, strongly values authentic, everyday objects in the classrooms (Folque & Siraj-Blatchford, 2011), but the ECE centre had mainly make-believe items in the play areas, except for some literacy-related items, like real empty packages in the grocery store. Could mathematical learning also be supported by playing with real, daily life objects? What knowledge about them would the children reveal?

The group was cheerful and energetic. Diverse in age (from three to six years old), socio-economic background, interests, and experiences, the children enjoyed coming together for storytelling and singing, and were used to free play in the afternoon. Isabel, the teacher, is a serene professional who promotes a quiet, respectful environment. The teacher-led activities were not planned for specific content areas, but there was a regular evaluation of the opportunities promoted for each area. In the school year, there had not been any activity directed at teaching measurement, but on some occasions, there had been the need to measure – for example, for wrapping presents or making collages with different clippings.

The room, large and well lit, was equipped for play by presenting play areas, or corners: doll's house, grocery store, library, games area, and construction area. The materials from the last two could be used on the large tables or in the reunion area (meeting place usually sitting on the floor). The play areas did not change much throughout the year and children were not invited to make suggestions regarding space and materials.

The teacher did not follow a specific pedagogical grammar (Oliveira-Formosinho, 2007), but valued play and children's self-directed activity, during part of the day.

In 2016, when the new National Curricular Guidelines for ECE were published, play became a strong theme for pedagogy (Ministério da Educação, 2016), but even before, Isabel already considered play to be central for children's experience of ECE. Her 30 years of experience had given her a strong sense of the potential of play for children's learning and well-being.

Placing the measuring tape in the doll's house

Based on the analysis of Christie's (2005) work on literacy-enriched play environments, it was decided to introduce a tailor's measuring tape into the doll's house, where most of the pretend play started. Among several measuring tools, the tape was identified as the most common in (some) children's homes and likely to have been experienced directly by the children, at least during their paediatric consultations. It was also an object that could be easily manipulated and moved by the children from the doll's house to anywhere in the room. Other tools, like the measuring jug, could be too restricted to the kitchen area and too narrow in terms of possible uses.

A large body of research has focused on the literacy-enriched play centre strategy in which play areas are stocked with reading and writing materials, and indicates that this type of manipulation of the physical environment is effective in increasing the range and amount of literacy behaviours during play, with positive learning impacts (Christie & Roskos, 2013). How children share literacy knowledge and process with one another in their play episodes has also been studied (Roskos & Christie, 2011). Play as a context for observing children's use of mathematical ideas is also valued in early childhood mathematics education, both in school and family care contexts (Gejard & Melander, 2018; Ginsburg et al., 2003; Hendershot et al., 2016; Wager & Parks, 2016).

The main point of discussion, and research interest, was how the measuring tape, a formal tool for standard measurement, would be introduced as a starting point for talking with children about measurement when the guidelines suggest that the formal standardised tool should be introduced after a trajectory of learning about measuring by comparing and using non-standard measures (Mendes & Delgado, 2008; Ministério da Educação, 2016). Would the children know what the object was? Ignore it? Use it for other purposes? A plan for helping the children notice the tape was devised in case the children didn't pick up the novelty in their doll's house.

The tape was placed in the doll's house without any special design. Because the room was usually cleaned after the day was over, it was not uncommon to find some changes in the disposition of materials or furniture.

Discussing the uses of the tape by the children

The situations that were observed and recorded showed three main uses of the tape: used as a different object not related to measurement, identified as a measuring tool but not manipulated as such, and used for play measuring (with no results). By placing the measuring tape, it was possible to introduce mathematical concepts in the child's play activities, as suggested by van Oers (1996).

Analysing the children's actions with Burghardt's criteria (2011) for recognising play reinforced the idea of playing with the measuring tape as a context for sharing ideas, experiences, and knowledge between the children, not as a measurement activity. This is particularly relevant when looking at four of the criteria:

1   the performance of the behaviour was not fully functional in the form in which it was expressed – it does not mean purposeless or having a delayed benefit but that it was not of immediate use as nothing was really measured nor was there a need to measure;

2   the behaviour was spontaneous, voluntary, intentional and pleasurable, as no instruction or suggestion was given to the children;

3   the actions of the two groups that identified the measuring tape differed from the functional expression of measuring because they were incomplete; and

4   it was initiated while children were with high levels of emotional well-being and it was uncoerced.

After the three days of observation, small conversations with the children were initiated by the teacher about the measuring tape. This was intended to start the discussion and reflection about the actions that children had been involved in, from a mathematical point of view, and arose from their activity (Björklund, Magnusson, & Palmér, 2019; van Oers, 1996).

All the children knew the name of the object and associated it with measuring. Half the children were able to identify situations where they remembered seeing the object being used: the large majority (65%) mentioned a family context, with family members as users. In the examples of reasons for measuring, this predominance of the family context is also present: measuring beds and furniture when moving, measuring for clothes sizes, measuring walls before painting them and deciding if a bag would fit into the trunk of the car were shared by the children. Still, half the children gave tautological answers: we measure to measure things.

When asked about their play with the measuring tape at the preschool, twelve children had stories to tell. Five of them shared uses that were not about measuring: the cowboy chasing the bandits, a yo-yo, strapping colleagues with the "ribbon." and even drawing shapes on top of the table – a mathematical activity not related to measuring. Eight children had measuring-related activities to tell about: measure the baby, the wall, the boards, the floor carpet, and more generically, "things."

In the conversations about the measuring tape experience, the children talked about length, height, and size, but not about distance, not even in terms of near or far. Their experiences were restricted to the family context with no mention of professional contexts, like doctor or seamstress. The examples that were mentioned suggest that some of the measurements were made with a different tape, not like the one that was presented (a tailor's measuring tape, that is, a ribbon of cloth or flexible plastic), but instead with a self-retracting metal tape, normally used during moving and construction work. Still, the children associated the two.

Learning from the tailor's measuring tape in the doll's house – the teachers

This experience was rich in uncovering layers of possibility and new questions about play and mathematics for the teachers participating in it. A complex object that represents the formal, standardised measurement activity of adults was placed in the play area for a group of children that had few experiences with measurement in the preschool. Still, the uses and knowledge revealed by some children were sophisticated and mathematically relevant. Also, observing and recording the way the children played allowed a meaningful conversation with the children about their experiences outside the school about a specific topic.

For a context that already valued play, the experience brought a renewed interest in ways of provoking play through the setting of the play areas for particular content areas. It also made clear the need to explore further, systematise, and expand the ideas the children had about measurement. It was common for Isabel to participate in children's play when invited, but there was little experience with making use of occurring mathematical phenomena (van Oers, 1996) or extending children's encounters with mathematics (Wager & Parks, 2016). Björklund et al. (2019) provide lines of action for teaching mathematics in play – confirming the direction of interest; providing strategies; situating known concepts; and challenging concept meaning – which have been valuable.

There was a discussion around the question whether what the children were doing and saying was mathematics. This discussion was immensely relevant and led to interesting readings (van Oers, 2010). The children's use of words related to measurement, the way they assigned meaning to the object and some actions, the connection between those actions and purposes or reasons for measuring, as well as the connection to numbers were analysed as mathematically relevant. But following up on the knowledge the children revealed in a mathematical way includes using the mathematical names but also its implications so that the meaning assigned includes and is clear about the mathematical connection. This goes beyond clarification and expansion. It was also important to realise, through Bishop (1988), how restricted the teachers' own view can be on what is mathematical in children's play. This leads to having many actions that are not taken as mathematical forms by the teachers and therefore do not gain mathematical meaning.

Commentary 3

Author: Lynne Zhang

Teacher in observation, design, and reflection

The vignette displayed professional methods, using clear records, background description, process, and language. It showed that the Portuguese teachers have solid foundations in observing and recording. Child–initiated free play provides many opportunities for teachers to observe what children already know, how they learn, and how they interact. Teachers can facilitate the next steps for the child, through modelling and using language. Observation of children's behaviour needs to be specific and needs to reflect children's behaviour, language, and expression (Curtis & Carter, 2002).

The learning story approach (Carr, 2001; Carr & Lee, 2019), which is close to the approach used to report the vignette in this chapter, is also widely accepted by Chinese teachers as an assessment method which can record children's learning and development by continuous description of children's behaviour in a real context. In the vignette reported in this chapter, such an approach not only describes the experience but also analyses it in terms of mathematics learning and possible future developments. In the following commentary, comparisons are made with an approach emanating from a specific Chinese programme which has been identified as relevant in terms of learning experiences for children.

Agency and scaffolding learning

Acknowledgement of and reflection on children's agency in play are important (Taguma, Litjens, & Makowiecki, 2012). Within the vignette, the children have the time and opportunity to create many forms of play with the tape, including using it as a "cowboy rope" as well as its social function of measuring length. Different children see different possibilities in their experiences with the tape, and the children are free to display their agency. This reflects a strong recognition that children should be encouraged to "be a child" and "learn through play" (especially child-initiated play).

Recognising agency gives children opportunities to undertake adventures, be inquisitive, and interact socially. All of these are very important for young children. Through their play interactions, children also develop language, logic, critical thinking, social skills, and mathematical knowledge and skills. This learning and development does not occur in a vacuum. "Knowing others" can support it.

In the "measuring tape" vignette, there are many opportunities for the children to direct and control their experiences. Concepts, meanings, processes, and values are being shaped and these belong to the learner. They are developed by the learner, constructed by the learner, owned by the learner, and shaped by the learner. They are shaped in response to messages received not just from the children and the teacher but from the whole environment, both physical and social.

The physical environment of early childhood classrooms and spaces is very important to suggest and support learning opportunities. Play areas with different objects and equipment create those opportunities. The organisation of the areas and the possible relations between them is also important and is part of the Portuguese guidance for early childhood education (Ministério da Educação, 2016). The vignette shows how small changes in the environment can enrich the children's experience. A simple, everyday object, combined with the playing dynamics that occurred in the space, promoted several explorations that also show the relevance of the social environment. Again, the Portuguese guidance document gives special importance to the way the group and the relationships are organised (Ministério da Educação, 2016), which allowed for the exchanges between the children and the way they respect and learn from each other about the measuring tape. The idea of intervening in the environment to continue challenging and supporting the learners is very present in the vignette. The learner is a "creative and adaptive" organism (Bishop, 1997, p. 126) and displays this when given the opportunity to do so.

While children sometimes learn from teachers' demonstrations and coaching, Vygotsky's theory of scaffolding points out that children often learn more working side by side with friends who are operating in their zone of proximal development (Mooney, 2013). Peer interactions are critical to a child's development of social and moral feelings, values, and social and intellectual competence (Piaget, 1965). Reading the Portuguese vignette from a Chinese perspective led to a discussion about the possibility of Chinese children learning mathematics through interactions with other children and teachers while playing, just as it has been illustrated in the vignette.

The age of artificial intelligence has required Chinese educators, families, and policymakers to ask what the purpose of early education is and how this can be supported through appropriate curriculum. The *Early Learning and Development Guidelines for Children Aged 3 to 6 Years* (Ministry of Education of the People's Republic of China, 2012) provides some answers. While the pedagogy of play and child-centred curriculum and traditional culture are seen as important, the Guidelines encourage children to develop thinking and problem-solving skills based on real-life experiences. In preschools, mathematics activities are designed to provide materials for children to handle and play with so that they can experience the importance and fun of mathematics according to their interests and knowledge.

Mathematics is an important part of Chinese culture. Children play with traditional developmental toys and games and work with cultural materials in order to broaden their mathematical knowledge and skills and understand the interrelationship between mathematics and other aspects of life. Chen Heqin's educational maxim of "big nature, great society" are used for reference to this approach.

Measurement activities in Happy Future Connection Developing Community

Measuring is concerned predominantly with comparing things according to a shared quality, and develops through comparisons using convenient, non-standard units, standardised units, and systems of units (Bishop, 1997). In every country, including China, there are many non-standard units as well as standardised ones for children to experience in early childhood education. The measuring tape central to the vignette reported in this chapter is just one sort of instrument carrying one sort of standardised unit.

The Happy Future Connection Developing Community programme (HFC DCP) is a new early developing community programme in Beijing based on the following principles:

- unity, connections and contextualisation
- transdisciplinary, formal and informal learning
- creative thinking and problem-resolving in real life context
- sensory based and integrated approach
- wisdom of traditional culture and future-oriented

The ecological model of human development (Bronfenbrenner, 1979) and the Chinese philosophy of harmony of people and earth are the main theoretical bases for HFC DCP, with its emphasis on environmental factors (and interactions among these factors) which influence, and are influenced by, the child. The ecological model provides the opportunity to address the future of the child through preparation for a world not yet seen, through a focus on thinking, not just facts and skills.

Measurement experiences in HFC DCP are designed as part of the children's daily routine and are often illustrated through traditional rhymes. Picture books are an ideal focus for commencing and continuing mathematical conversations.

Bishop (1997) demonstrates that mathematics is a universal language that is used across many aspects of children's lives including art, music, architecture, natural and social sciences, and technology. There are many opportunities in all of these fields to explore and compare size, shapes, angles, and spatial relationships, and to enhance knowledge, skills, and dispositions in measurement.

Each year a *Let's measure week* is organised where a variety of measuring instruments, both traditional and modern, are available for children and families. In terms of the measurement of time, the 24 Solar Terms project – an ancient Chinese guide to annual weather (China Highlights, 2019) – is helpful to experience the natural change of the seasons and the weather. Children, families, and educators keep track of the light and dark together and begin to develop their knowledge of the future from their sense of time (Seginer, 2009).

Many different roles are played in the educational system by people who share the responsibility for the mathematical enculturation process – teachers, teacher educators, advisers, inspectors, curriculum developers, resource providers, researchers (Bishop, 1997). Families and children also play important roles. Both in the Chinese example of HFC DCP and the Portuguese approach illustrated by the vignette reported in this chapter, active engagement, interaction, opportunities, and reflection play their part in the development of important mathematical ideas.

The articulation between encountering the concepts, experience, and purpose of measuring in picture books and real-life experiences, which may be prompted by objects like the measuring tape, provide context and connections that support children's learning. The Portuguese vignette is an example of how much knowledge and daily life experiences children bring to their learning experiences in early childhood education. Finding the right environments, challenges, and social interactions to expand those experiences is the role of any early childhood education programme.

References

Bishop, A. J. (1988). Mathematics education in its cultural context. *Educational Studies in Mathematics, 19*(2), 179–191.

Bishop, A. J. (1997). *Mathematical enculturation: a cultural perspective on mathematics education.* Mathematics Education Library, volume 6. Dordrecht: Kluwer Academic Publishers.

Bishop, A. J. (2016). Can values awareness help teachers and parents transition preschool learners into mathematics learning? In T. Meaney, O. Helenius, M. L. Johansson, T. Lange, & A. Wernberg (Eds.), *Mathematics education in the early years* (pp. 43–56). Springer.

Björklund, C., Magnusson, M., & Palmér, H. (2019). Teachers' involvement in children's mathematizing – beyond dichotomization between play and teaching. *European Early Childhood Education Research Journal, 26*(4), 469–480. https://doi.org/10.1080/1350293X.2018.1487162

Boavida, A. M., Paiva, A., Cebola, G., Vale, I., & Pimentel, T. (2008). *A experiência matemática no Ensino Básico: Programa de Formação Contínua em Matemática para Professores dos 1.º e 2.º Ciclos do Ensino Básico* [The mathematical experience in basic education: Programme of Continuing Teacher Education in Mathematics for 1st and 2nd Cycle of Basic Education]. Lisbon: DGIDC/ME.

Bronfenbrenner, U. (1979). *The ecology of human development.* Cambridge, MA: Harvard University Press.

Burghardt, G. M. (2011). Defining and recognizing play. In A. Pellegrini (Ed.), *The Oxford handbook of the development of play* (pp. 9–18). New York: Oxford University Press.

Buys, K., & de Moor, E. (2008). Domain description measurement. In M. van den Heuvel-Panhuizen (Ed.), *Young children learn measurement and geometry: A learning-teaching trajectory with intermediate attainment targets for the lower grades in primary school* (pp. 15–36). Rotterdam: Brill.

Carr, M. (2001). *Assessment in early childhood settings: learning stories.* London: SAGE.

Carr, M., & Lee, W. (2019). *Learning stories in practice.* London: SAGE.

China Highlights (2019). *The 24 solar terms.* Retrieved from www.chinahighlights.com/festivals/the-24-solar-terms.htm

Christie, J. (2005). Literacia – Contextos lúdicos enriquecidos [Literacy-enriched play environments]. In C. Neto (Ed.), *Jogo e desenvolvimento da criança* [Play and children's development] (pp. 140–150). Lisbon: Faculdade de Motricidade Humana.

Christie, J., & Roskos, K. A. (2013). Play's potential in early literacy development. In R. E. Tremblay, R. G. Barr, R. Peters, & M. Boivin (Eds.), *Encyclopedia on early childhood development* (pp. 1–6). Montreal: Centre for Excellence for Early Childhood Development. Retrieved from www.child-encyclopedia.com/sites/default/files/textes-experts/en/774/plays-potential-in-early-literacy-development.pdf

Curtis, D., & Carter, M. (2002). *The art of awareness: how observation can transform your teaching.* St. Paul, MN: Redleaf.

Folque, M. A., & Siraj-Blatchford, I. (2011). Fostering communities of learning in two Portuguese preschool classrooms applying the Movimento da Escola Moderna (MEM) Pedagogy. *International Journal of Early Childhood, 43*(3), 227–244.

Gejard, G., & Melander, H. (2018). Mathematizing in preschool: children's participation in geometrical discourse. *European Early Childhood Education Research Journal, 26*(4), 495–511. https://doi.org/10.1080/1350293X.2018.1487143

Ginsburg, H. P., Lin, C., Ness, D., & Seo, K.-H. (2003). Young American and Chinese children's everyday mathematical activity. *Mathematical Thinking and Learning, 5*(4), 235–258. https://doi.org/10.1207/S15327833MTL0504_01

Hendershot, S. M., Austin, A. B., Blevins-Knabe, B., & Ota, C. (2016). Young children's mathematics references during free play in family childcare settings. *Early Child Development and Care, 186*(7), 1126–1141. https://doi.org/10.1080/03004430.2015.1077819

Marchese, M. (2016). *O contributo de contextos lúdicos enriquecidos para o desenvolvimento de aprendizagens matemáticas na educação pré-escolar no âmbito da medida de comprimento* [The contribution of enriched play environments for mathematical learning in early childhood education – length measurement] [Master's thesis, Escola Superior de Educação de Viseu, Instituto Politécnico de Viseu]. Retrieved from http://repositorio.ipv.pt/handle/10400.19/4905

Mendes, M., & Delgado, C. (2008). *Geometria* [Geometry]. Lisbon: DGIDC/ME.

Ministério da Educação. (2016). *Orientações curriculares para a educação pré-escolar* [Curricular guidelines for early childhood education]. Lisbon: Ministério da Educação.

Ministry of Education of the People's Republic of China. (2012). *Early learning and development guidelines for children aged 3 to 6 years*. Retrieved from www.unicef.cn/sites/unicef.org.china/files/2018-10/2012-national-early-learning-development-guidelines.pdf

Monteiro, L. M. (2012). *A Medida na Educação Pré-Escolar: Um estudo centrado em experiências integradoras* [Measurement in preschool education: a study Centred on integrative experiences] [Master's thesis, Escola Superior de Educação de Viana do Castelo, Viana do Castelo].

Moreira, J. V. (2018). *Materiais não estruturados na Geometria e Medida em EPE e no 1.º CEB* [Non-structured materials in geometry and measurement in early childhood and primary education] [Master's thesis, Escola Superior de Educação de Paula Frassinetti].

Moreira, M., & Oliveira, I. (2003). *Iniciação à Matemática no Jardim de Infância* [Introduction to mathematics in the kindergarten]. Lisbon: Universidade Aberta.

Mooney, C. G. (2013). *Theories of childhood: an introduction to Dewey, Montessori, Erikson, Piaget, and Vygotsky.* St Paul, MN: Redleaf.

Oliveira-Formosinho, J. (2007). Pedagogia(s) da infância: reconstruindo uma práxis de participação [Childhood pedagogy(ies): reconstructing a participatory praxis]. In J. Oliveira-Formosinho, T. Kishimoto, & M. Pinazza (Eds.), *Pedagogia(s) da infância. Dialogando com o passado, construindo o futuro* [Childhood pedagogy(ies). Dialogues with the past, building the future] (pp. 13–36). São Paulo: ArtMed.

Piaget, J. (1965). *The moral judgment of the child.* New York: Free Press.

Ponte, J. P., & Serrazina, L. (2000). *Didática da Matemática do 1.º Ciclo* [Didactics of mathematics for primary education]. Lisbon: Universidade Aberta.

Rodrigues, M. (2010). *O sentido de número: Uma experiência de aprendizagem e desenvolvimento no pré-escolar* [Number sense: An experience of learning and development in preschool education] [Doctoral thesis, Faculdad de Educación da Universidad de Extremadura], Badajoz, Spain.

Roskos, K., & Christie, J. (2011). The play-literacy nexus and the importance of evidence-based techniques in the classroom. *American Journal of Play, 4*(2), 204–224.

Seginer, R. (2009). *Future orientation: Developmental and ecological perspectives* (6th ed.) New York: Springer.

Silva, J. C. (2019). *Recomendações para a melhoria das aprendizagens dos alunos em Matemática* [Recommendations for the improvement of students' learning in mathematics]. Report from the Commission on Mathematics Teaching. Lisbon.

Taguma, M., Litjens, I., & Makowiecki, K. (2012). *Quality matters in early childhood education and care: Portugal 2012.* Paris: OECD.

Tyminski, A., Weilbacher, M., Lenburg, N., & Brown, C. (2008). Ladybug lengths: Beginning measurements. *Teaching Children Mathematics, 15*(1), 34–37.

van den Heuvel-Panhuizen, M., & Buys, K. (2005). *Young children learn measurement and geometry (TAL Project)*. Utrecht: Freudenthal Institute.

van Oers, B. (1996). Are you sure? Stimulating mathematical thinking during young children's play. *European Early Childhood Education Research Journal, 4*(1), 71–87.

van Oers, B. (2010). Emergent mathematical thinking in the context of play. *Educational Studies in Mathematics, 74*(1), 23–37. https://doi.org/10.1007/s10649-009-9225-x

Wager, A. A., & Parks, A. N. (2016). Assessing early number learning in play. *ZDM, 48*(7), 991–1002. https://doi.org/10.1007/s11858-016-0806-8

PART 4

Locating

7

FANTASTIC MR FOX

Anne Hj. Nakken, Camilla N. Justnes, Oda Bjerknes,
and Simone Dunekacke

Vignette

"This time we must go in a very special direction," said Mr Fox, pointing sideways and downward.

So he and his four children started to dig once again.

(Dahl, 1970, p. 33)

This vignette is an observation of three play sequences starting with a free play activity initiated by children in their outdoor environment. In Norway, there is a culture of seeing children's free play as a possible starting point for exploratory activities. The *Framework Plan for Kindergartens – Content and Tasks* (Ministry of Education and Research, 2017) states explicitly that staff shall create opportunities for mathematical experiences by enriching children's play with mathematical ideas and in-depth conversations. Before the observation, the children had worked on a project about animals in the forest, including the red fox (Vulpes vulpes). As a part of this project, the children had listened to the story *The Fantastic Mr Fox* by Roald Dahl (1970).

Mr Fox is most certainly an expert in locating, as he digs tunnels in different directions to outmanoeuvre the three nasty farmers. No wonder the children in this vignette are intrigued and inspired by this story. The children were, both during and after the project, occupied by foxes and fox burrows in their free play.

In this vignette, we have emphasised how the children's initiative is heard and supported with careful teacher participation. The pedagogical intention is to enrich the children's play with more mathematical experiences, like the Framework Plan states, but without disturbing or changing the children's play too much. By carefully observing the children's actions and ideas, and analysing the mathematical possibilities in their play, the pedagogue poses comments, questions, and material to support the children's further exploration and mathematical development.

Three children, Julia, Samuel, and Norah, play in the sandbox. They are four and five years old. The pedagogue, Emma, observes them. Julia is digging a fox burrow in the sand, and Samuel and Norah join her. Julia asks them if they can help her to dig deeper. Samuel points out that the burrow only has one opening, whereas in a proper burrow there are many more. Samuel says that the fox needs to go in and out several places, as Mr Fox did in the story when the farmers closed one of the openings and tried to catch him. Julia agrees, and she starts to dig more openings (Figure 7.1). Norah fetches plastic animals to live in the burrow. The three of them try to play with the animals in and out of the burrow, but they experience that it easily collapses. Emma recognises their rising tension and frustration in the sandbox.

Emma wants to support the continuance of the play and offers them an idea to strengthen their construction. She says, "Maybe I can bring you

FIGURE 7.1 Digging a burrow

a box to make your burrow stronger?" The children hesitate and do not seem convinced by Emma's idea, but she brings a shoebox and a wallpaper knife to the sandbox anyway. Emma places the shoebox in the deep hole in the sand. She asks the children to help her cover it with sand on the top. Julia agrees to help her, but Samuel and Norah just sit quietly, observing them. Emma uses the wallpaper knife and makes a hole in one of the sides. "There," Emma says, "maybe this can be the opening?" Julia puts her plastic fox in the opening. Samuel and Norah join in, with both putting more sand on the box and with their plastic animals. After a while, Norah points out that this burrow also should have more openings. She asks Emma if she can make another opening on the opposite side of the box. The other children suggest making more openings in different places, and Emma helps them.

The children put their plastic animals in the first opening, then put their arm in another opening and pull the animal out. They do this several times with different openings, but not systematically. Emma asks, "How many different routes can the animals take?" The children try to explain and show their thinking, but they find it difficult. It seems like the box in the sand limits them because of tight openings, clothes, arms, and because they cannot see through the box. The children are a bit frustrated and lose interest in exploring further since they cannot use the box to show their thinking.

Emma suggests that the children can explore different pathways through a fox burrow in the jungle gym, pretending that the gym is the burrow. The children latch onto the idea and immediately start to explore the various possibilities for crawling in and out of the gym (Figure 7.2). They play and run through the burrow over and over again. Emma repeats her question from the sandbox, "How many different routes can the animals take through the burrow?" The children show different pathways by crawling through the gym, they shout out their route, and point where they have been crawling, "I went this way", "Look at me! I found another way." They sometimes disagree if a pathway is new or if someone already found it. Emma asks, "At what point must the fox turn or change direction to make a new route?" The children all crawl in the same opening and stop at the first "crossroad" inside the jungle gym and delegate different directions to crawl from there. To support them in keeping track of their different routes, Emma asks them, "How can we keep track of our different routes?" Samuel tries to count, but he loses track of the different routes. Julia suggests that they sit in front of the opening they have used to close it for others to crawl into.

Emma gives the children ropes to bring with them when they crawl. She says, "Now, you can use these ropes to see where you have been or not." The children take one rope each and crawl through the gym in turns. After a

FIGURE 7.2 Jungle gym

while, Emma invites the children to make sure that none of the ropes shows the same pathway and to count all of the ropes. Norah says, "There are 6 ropes, so there must be 6 routes through the burrow." Samuel says, "But the foxes are more clever! They don't need ropes!" Julia agrees, "Yes, that's smart! If the foxes used ropes the farmers would find them." They all agree that the foxes are clever.

Reflective questions

1   When does Emma *not* facilitate engagement?
2   What would you have done in this situation?
3   Which competence does Emma use when she changes her suggestions based on the children's response?

Commentary 1

Authors: Anne Hj. Nakken and Camilla N. Justnes

Early childhood mathematics education in Norway

The Norwegian preschool practice is situated in a social pedagogical tradition where childhood has an intrinsic value. The preschools promote and practice a holistic approach to children's development by intertwining care, play, and learning. Children's freedom to explore their surroundings, including those of mathematics, is highly regarded. The pedagogical practice of preschools is described in the *Framework Plan for Kindergartens – Content and Tasks* (Ministry of Education and Research, 2017). In this plan, the mathematics domain has its own chapter, "Quantities, spaces and shapes," which explicitly states expectations of what the institution and the pedagogues should provide regarding mathematical opportunities in the preschool environment. The Framework Plan focuses on children's opportunities to experiment, discover, and play, which according to Devlin (2012) brings mathematics in preschool close to the work done by mathematicians. Although not formally acknowledged, the mathematical objectives highlighted in the Framework Plan can be traced back to Bishop's six mathematical activities (Bishop, 1988b). Bishop's universal activities remind us of the broad array of mathematical competencies children develop in their early years. The different competencies together constitute deep mathematical understanding. Locating is a part of these competencies. Bishop describes locating as exploring one's spatial environment. Spatial reasoning enables children to understand both their spatial environment and other mathematical topics (Nguyen et al., 2016).

Noticing the mathematical potential and enriching play

In the Norwegian preschool context, children's competencies are expressed through play and daily life activities. However, while the potential for play to contribute to children's mathematical understanding has long been recognised, this potential is only realised if the mathematics in play is noticed, explored, and talked about (Dockett & Goff, 2013). In the vignette, we can see that Emma notices the mathematical potential in the children's play and tries to facilitate further exploration and dialogue in different ways. Her mathematical knowledge, pedagogical knowledge, understandings, beliefs, and perceptions influence the choices she makes (Ball, Thames, & Phelps, 2008; Cooke & Bruns, 2018). She sees the potential for the children to have valuable spatial experiences in their play and supports spatial experiences. She chooses to bring a shoebox into the sandbox. Emma intends to make the construction more solid, inviting the children to use and explore spatial movements and spatial language further. Emma listens and observes with genuine interest, and she helps the children when they ask her to make more openings in the box. Later, Emma decides to invite the children to move the exploration from the sandbox to the jungle gym. By doing this, she makes it possible for them to create new strategies to solve the problem by moving the exploration to a more suitable model. Emma supports the children's enjoyment in the process of developing their locating skills, without introducing the concepts to them or offering them her own solution. By inviting the children to demonstrate, model, explain, justify, transfer, connect, and describe, teachers expand children's mathematical thinking in an effective way (Cheeseman, 2009).

Emma recognises and responds to the children's initiative in the situation, and she enriches their play with mathematics. Björklund, Magnusson, and Palmér (2019) suggest four ways the pedagogue can enrich children's play: confirm the children's direction of interest, provide strategies for problem-solving, situate known concepts, and challenge the meaning of the concepts. In the vignette, we can see traces of all four of these. Emma confirms the children's interest in the movement through the burrow and builds on this when she challenges the children to find all the different paths through the burrow and provides an opportunity for them to develop their own strategies. During the entire situation, Emma situates the concept actively, and she gives the children various opportunities to expand their views on what "through" really means. Her actions are an important part of her professional pedagogical practice related to mathematics, as there is a significant relation between teachers "maths talk," and the growth of children's maths skills (Klibanoff et al., 2006).

Problem-solving

Emma invites the children to think mathematically by asking questions that have the potential to foster excitement and curiosity, and challenge them to solve problems. Problem-solving of realistic everyday challenges and logical reasoning related to this, are found as powerful facilitators of mathematical learning in the early years (Perry & Dockett, 2005). According to Dewey (1934), it is essential that teachers build on children's past experiences when inviting them into activities where they are challenged to "find out." In the vignette it is clear that Emma's first attempts to foster problem-solving by asking questions is not fruitful. However, by staying with the children and being patient, Emma manages to find a task that activates them. By moving the investigation from the shoebox to the jungle gym, the children get engaged, and Emma guides the children's attention to previously unseen features of the problem. She makes it possible for the children to find and show their solutions by crawling through the gym, gaining a new perspective on what is being asked. In order to become owners of and active participants in problem-solving, the children must be invited into a landscape of investigation (Alrø & Skovsmose, 2004). In the vignette, we see that the children accept the invitation by Emma and they locate themselves in the activity. Their reasons for accepting her invitation could be their interest in and familiarity with the burrow, relationships with the other children and/or Emma, or a wish to cooperate with someone. In the course of the investigation, the children sometimes disagree if a pathway is new or not. Emma clears up this disagreement not by taking over the thinking or giving the children the answer, but by asking another question, "How can we keep track of the different routes?" By posing a new question, she calls for the children's further thinking and generates discussion among them. She stimulates their views of themselves as active thinkers and problem-solvers. This harmonises with Dewey's (1934) attention to the importance of the children's reflection on what they are discovering to expand their understanding.

Locating

Locating includes both spatial orientation and spatial visualisation (Bishop, 1988b). Spatial orientation involves understanding and operating on relationships between different positions in space, first with respect to one's own position and one's own movement through space, then towards an abstract perspective that includes maps and coordinates. Spatial visualisation includes creation, interpretation, use, and reflection on images in our minds with the purpose of thinking, communicating, and developing ideas and understanding. In the vignette, the children's ability in both spatial orientation and spatial visualisation is stimulated. When the children describe and show the positions and movements of the fox in the sandbox and in the jungle gym, they use their spatial orientation sense by connecting paths between the openings. When challenged to find out how many different routes the fox can take, the children must use and reflect on their visual mental image of the burrow. Spatial sense and problem-solving are strongly connected to each other as it is important to be able to visualise and use images when reasoning (Boaler et al., 2016; Wheatley, 1997).

Representations

Mathematical ideas can be represented in different ways, and these are connected to each other. The representation of space is built up from active manipulation of the environment (Piaget & Inhelder, 1948). Bishop (1988b) describes locating as exploring, conceptualising, and symbolising one's spatial environment with representations like models, diagrams, drawings, words, or other means. Clements and Sarama (2009) remind us that children develop spatial skills through bodily and sensory experiences. This importance of bodily experiences of space is also reflected in The Norwegian Framework Plan as it declares that "Kindergartens shall enable the children to use their bodies and senses to develop spatial awareness" (Ministry of Education and Research, 2017, p. 53). Bruner (1966) described children's mathematical development as a loose progression from an action-based (enactive) representation, via a visual (iconic) representation, to a symbolic representation. The children in the vignette experience the translation between the iconic representation and the enactive representation when they hear about the fox and look at pictures in the book, and translate this to themselves acting out the fox investigating the different routes bodily through the burrow.

In addition, they meet different models of the burrow: the sandpit, the shoebox, and the jungle gym. Later they represent the different routes through the burrow with oral language, with their bodies, and with ropes. The children's experiences with different representations and their connections will lay the foundation for a deep understanding of spatial relationships in verbal language, pictures, drawings, or symbolic maps later.

Summary

As we have seen in the vignette, locating is a complex mathematical activity which consists of many sub-skills. Understanding spatial relationships includes using and understanding spatial language, to navigate in varied environments, to make and manipulate mental images, to predict what will happen, and to solve problems. During planned activities and children's free play in preschool, opportunities for meaningful mathematical exploration and discussions will arise. However, the teachers must see and build on this potential by highlighting ideas and relationships. By actively engaging children in thinking and communicating, they will participate in the joint construction of knowledge. This vignette reminds us to conceptualise a broad view of mathematics where locating has a central and natural place, and where children and teachers together can explore mathematical ideas with great interest.

Commentary 2

Author: Oda Bjerknes

Introduction

The vignette about the children acting out *Fantastic Mr Fox* describes a typical situation in Norwegian preschools. Children convert stories they have been told, and which have captured their imagination, into free play. Free play belongs to children and is characterised as fun, voluntary, and spontaneous. In Norwegian preschools, there is still discussion about free play, especially concerning the role of the pedagogical staff. Many interested in early childhood education believe that free play is children's domain and that staff should take a step back.

The teacher in this vignette, Emma, does not disengage. At first, she allows the children to play independently while observing them. Once Emma sees that the children are starting to become frustrated, and tension begins to rise, she steps in and offers support and guidance. Emma demonstrates her competencies in maths by spotting an opportunity for the children to practise spatial awareness while they play.

Guiding the children towards new experiences

Emma uses her knowledge and experience of children's play and development by offering a shoebox to the children playing in the sandpit. She knows it will reinforce the children's building project. When Emma intervenes to offer guidance and expand the scope of the children's play, she takes an active part. The children are able to continue their play because Emma has enriched it. Emma makes further decisions that influence the children's play. At times she leads, causing some of the children to hesitate a bit, but soon they continue to build on her initiative.

If we look again at what actually constitutes free play, I understand what is described in the vignette as fun and voluntary, but I would no longer call it free play after Emma has taken the lead. I view Emma's role in this vignette as a type of supportive scaffolding – a metaphor developed by Jerome Bruner based on Vygotsky's theories (Bruner, 1978). Vygotsky believed that learning happens in interaction and dialogue with someone who is more knowledgeable about the concept the children are learning about. With support from someone more knowledgeable, a child can extend their zone of proximal development. Emma builds scaffolding around the situation by supporting and expanding when necessary and steps back when the play continues to flow again, yet she remains at hand for further support.

Mathematics and language are linked

Emma notes that the children are exploring potential routes for the plastic animals to travel through the burrow. She also notices that they are not being very systematic. She, therefore, asks them, "How many different routes can the animals take?" The question could encourage them to be more systematic in the way they explore the routes, and it could create an opportunity for problem-solving.

As a practitioner, I often find it difficult to ask good questions, questions that attract the children's attention and motivate them. I prefer to avoid questions which encourage the children to say what they think I want to hear. I have found that a great deal of reflection and practise is needed before the questions start to come naturally. In the vignette, the children try to explain and demonstrate their thinking in response to Emma's questions. It is likely that the children used mathematical terminology in these conversations.

In the Norwegian preschool education system, mathematics is among the topics which are compulsory. The focus is on mathematics and its didactics, as well as how it must be viewed in the context of other subjects and pedagogy. The aforementioned system includes interdisciplinary perspectives. In the Norwegian preschool teacher education system, the subjects are classified into "areas of knowledge." Mathematics is a part of the area called "language, text, and mathematics" (Ministry of Education and Research, 2012, p. 4). Within the *Framework Plan for Kindergartens – Content and Tasks*, asking questions, reasoning, argumentation, and seeking solutions are important features in both mathematics and language.

Different ways of solving problems

The Norwegian Framework Plan recently introduced problem-solving as an element of mathematics in preschool. Preschool teachers should stimulate and support children's capacity for, and perseverance in, problem-solving (Ministry of Education and Research, 2017). Perseverance when working with a problem requires patience from both the children and the practitioner. If the children are to take an active part in the problem-solving process, the solution must not be given to them; rather, they must be allowed sufficient time to think about it and try things out. Problem-solving in preschools and schools allows children to develop all-round mathematical skills, which they will benefit from for the rest of their lives. Emma encourages the children to persevere whilst inspiring them to consider different solutions. Again, she takes a more proactive role and suggests that the children move across to the jungle gym. For young children, a physical approach is often necessary, to make an abstract problem more concrete and comprehensible. The Framework Plan states that by engaging with quantities, spaces, and shapes, the preschool should encourage the children to use their bodies and senses to develop spatial awareness (Ministry of Education and Research, 2017) – something they certainly do while playing in the jungle gym. Once the children get to the jungle gym and start exploring potential routes through it, Emma repeats her question from the sandpit. By reiterating the question, she allows the children to explore the same theme in a different setting.

The children identify a number of routes through the jungle gym, but they struggle to remember them all. Emma aids the children by asking questions such as, "At what point must the fox turn or change direction to create a new route?" The children discuss strategies but are unable to create abstract, mental maps. Emma appears to recognise this and offers the children some rope that they can use to mark out the routes. And once again, Emma builds scaffolding around the children's play. In this context, using physical objects during mathematical play is often a great help for young children.

Pedagogical awareness

In this story, Emma chose to steer the children towards mathematical play by focusing on spatial awareness. She could also have opted to ask the children other questions that would have given them different experiences. Preschool teachers make these choices every day. The questions we ask determine what happens next. My point is that the teacher's "teaching methods" require a high degree of pedagogical awareness because we do not teach in the traditional sense. Children exploring a fox burrow do not feel that they are having a maths lesson. They probably also find that it is their activity since they took the initiative and remain active and involved throughout. The Framework Plan places emphasis on child participation in everyday life, and most activities in preschool are interdisciplinary. In practice, this means that Norwegian preschool teachers need to be knowledgeable in a range of areas, and must plan for the unplanned. They must also be sensitive in their approach, and observant so that they can offer support in any situation. This requires the ability to switch between planned and spontaneous situations. As for the story, I can spot potential for exploring problem-solving and spatial awareness in more depth.

Summary

The Framework Plan states that preschool staff should encourage children to find pleasure in mathematics and take an interest in mathematical relationships based on their forms of expression (Ministry of Education and Research, 2017). This requires teachers to be knowledgeable about children's different forms of expression – in this case, play – and to build scaffolding and guide the children. The teacher in this vignette recognises that the children should make progress in preschool, and she knows that spatial awareness develops from actual corporal experiences into cognitive mental maps. The children in the vignette encounter mathematical challenges and are able to experience a joy of maths because Emma demonstrates her competency in the subject and spots an opportunity to integrate mathematics into play.

The children are able to make contributions and retain their curiosity and motivation for learning as Emma expands on their original initiative and improvises within this. Such improvisation highlights how preschool teaching differs from the idea of conventional teaching. The vignette is but one example of how a preschool teacher teaches by drawing on the children's own forms of expression and integrating mathematics into everyday activities. Emma knows what she wants to achieve, and her thoughtful decisions give the children new experiences and insights.

By enriching their play, and allowing them to explore using physical objects, Emma permits the children to discover that there can be many different approaches to a problem and a variety of approaches to mathematics in general. The children also identify how there are multiple solutions to a problem on account of the observant teacher who steps in to expand the situation and ask productive questions.

Commentary 3

Author: Simone Dunekacke

Mathematics education in German kindergarten

In contrast to other countries such as the U.S. or U.K., the German kindergarten focuses on a social-pedagogic approach (Diskowski, 2009), as does the Norwegian kindergarten. Learning should take place during everyday situations (like circle time, free play, or lunch) and should be induced by children's interests. Despite these foundations, the current debate is leaning more and more towards questions of domain-specific learning of children. The causes of this development are the results of the 2000 Programme for International Student Assessment (PISA) as well as an upcoming discourse about the (limited) quality of early childhood education and care, especially from the perspective of domain-specific child development. Empirical results indicate that a high-quality early childhood education is necessary for children's mathematical learning in particular (Ulferts & Anders, 2016). Children benefit from this support without forfeiting the development of social skills (Kluczniok et al., 2016). One of the main results of this discussion was the introduction of curricula, which are called *Bildungspläne*, for each German federal state (Diskowski, 2009). Most of them include mathematics as a relevant topic for kindergarten children. However, there are wide discrepancies in the quantitative and qualitative content concerning mathematics in the *Bildungspläne*, ranging from only a few sentences to detailed explanations of what should be done with the children.

Next to this discourse, (early) mathematics education research reflects an increasing interest in children's mathematical development before the start of school and ways to support this development (Peter-Koop & Scherer, 2012). This was visible in a number of recent studies focusing on children's development (Lüken, Peter-Koop, & Kohlhoff, 2014; Unterhauser & Gasteiger, 2018). Furthermore, there is an ongoing discussion about appropriate ways to support children's mathematical learning within kindergartens (for an overview: Schuler, 2013). German early childhood mathematics education addresses various mathematical contents (quantity, space and shape, units and measuring) as well as cognitive processes (mathematical communication, problem-solving) (Benz, Peter-Koop, & Grüßing, 2015). Many of the ideas from Bishop's (1988a) fundamental activities can be found in this distinction. Irrespective of the content, as well as the methodological way to foster children, two maxims of early mathematics education should be an orientation on the subject of mathematics and a child orientation (Gasteiger, 2015). Subject orientation means to offer children age-appropriate but mathematically correct content. This is important with respect to children's learning during the transition to primary school (Gasteiger, 2015). Child orientation indicates that mathematical learning should address children's ways of learning as well as their interests and skills (Gasteiger, 2015).

Preschool teachers have been identified as playing an important role in young children's mathematical learning (Gasteiger, 2015). Competence models, as presented in Figure 7.3, are used to describe preschool teacher competence. Competence models should first be seen as theoretical descriptions of the complex

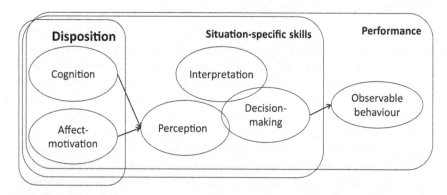

FIGURE 7.3 Model of professional competence (Blömeke, Gustafssonm, & Shavelson, 2015, p. 7)

construct of teacher competence. They allow an empirical investigation of competence, or aspects of competence. Based on this, it becomes possible to identify effective approaches for teacher training or further education. Figure 7.3 illustrates a model of preschool teacher competence. Cognition includes the different knowledge sources teachers need. Affect-motivational aspects can include, along with others, aspects of preschool teacher's attitudes. These dispositions can help teachers identify relevant learning situations (perception), connect them to mathematical learning (interpretation), and decide how to act (decision-making). These situation-specific skills result in the observable behaviour of the preschool teacher. Research has found that preschool teachers need knowledge of mathematical content and pedagogical content (Dunekacke et al., 2016; McCray & Chen, 2012). Next to knowledge, teachers need skills to recognise opportunities for mathematical learning in the open-ended situations of German kindergarten (perception), to interpret these situations from a maths-related perspective (interpretation), and to generate a maths-related action (decision-making) (Dunekacke et al., 2016; McCray & Chen, 2012). Next to these cognitive-grounded skills, preschool teachers' attitudes towards (early) mathematics are found to be important in fostering children's mathematical learning (Benz, 2012; Bruns, Eichen, & Gasteiger, 2017; Oppermann, Anders, & Hachfeld, 2016; Thiel, 2010).

In contrast to many other countries, the majority of preschool teachers in Germany are trained in post-secondary specialised colleges (*Fachschulen*). Even if domain-specific education has become much more in focus in preschool teacher training, there is only limited data available on the number of opportunities to learn early mathematics education in preschool teacher training. It is assumed that these domain-specific opportunities are not offered enough but would help prospective teachers generate more maths-related knowledge and skills (Blömeke, Dunekacke, & Jenßen, 2017; Strohmer & Mischo, 2015).

Relation to the vignette of Fantastic Mr Fox

The vignette of *Fantastic Mr Fox* gave great practical insight into these empirical findings of domain-specific professional competence of preschool teachers. What we see (observable behaviour) is Emma, the preschool teacher, interacting with Julia, Samuel, and Norah.

The vignette starts while Emma is observing the children; in other words, she perceives the children. From what is shown in the vignette, we may assume that Emma might interpret the situation focusing on the children's interests first (Emma recognises the frustration of the children). Next, she makes a decision (decision-making) to stimulate the children to solve the problem. Emma then goes on with her observation (perception) and realises that the children explore the different routes through the burrow and the different exits. What can be seen in the vignette is that Emma asks a concrete maths-related question ("How many different routes can the animals take?"). Emma might interpret the situation as a possibility to involve the children in mathematical reasoning. With the stimuli, Emma gave age-appropriate and child-oriented use of maths-related language, which also offers a high level of subject orientation (How many "different routes" instead of only "routes") (Gasteiger, 2015; Klibanoff et al., 2006). It might be possible that this is linked to Emma's content knowledge on the one hand (there are a limited number of ways through the burrow) as well as her pedagogical content knowledge (use technical language, offer concrete questions).

Emma then transfers the situation to the jungle gym. That can be seen as another decision-making process between these situations. She gives the children the possibility of working on the same question ("How many different routes can the animals take?") but offers another mode of representation. Exploring the routes in the jungle gym burrow is very action related, much more than exploring the routes in the burrow in the sandbox, which are invisible in parts.

If we assume that Emma's decision was based on her pedagogical content knowledge (children should learn in different situations and representation modes), this part of the vignette might gave another concrete insight into the relationship between teacher's knowledge and the skills to perform in open-ended kindergarten situations. The children then decide on a strategy to solve the problem (count the routes by themselves) but struggle, because it is unclear where a new route starts. Emma observes (perception) this and recognises (interprets) the struggle of the children. She links the attention of the children to the mathematical content ("at what point must the fox turn or change direction to make a new route?") and offers a possible insight into her content knowledge (crossings are important for the number of routes). Furthermore, she offers the children help to develop an appropriate strategy to solve the mathematical task by given them ropes to mark the routes.

Summary

In summary, the vignette gave a nice insight into early childhood mathematics education. It allows identifying elements of teachers' professional competence and their importance for offering high-quality, subject-oriented as well as child-oriented (Gasteiger, 2015) learning situations. The vignette also offers a great example of how themes can lead through different educational domains (language, natural science, and other) and with that foster different kinds of skills in children (Kluczniok et al., 2016). However, this insight is limited to what is visible within the vignette. All assumptions about the activated knowledge and skills can only be seen as (maybe plausible) considerations. To validate the considerations, more insight into the situation or other situations of mathematical learning would be necessary, for example, by further observations or interviews with Emma.

References

Alrø, H., & Skovsmose, O. (2004). *Dialogue and learning in mathematics education*. Dordrecht: Kluwer Academic Publishers.

Ball, D. L., Thames, M. H., & Phelps, G. (2008). Content knowledge for teaching: what makes it special? *Journal of Teacher Education, 59*(5). https://doi.org/10.1177/0022487108324554

Benz, C. (2012). Attitudes of kindergarten educators about math. *Journal für Mathematik-Didaktik, 33*(2), 203–232. https://doi.org/10.1007/s13138-012-0037-7

Benz, C., Peter-Koop, A., & Grüßing, M. (2015). *Frühe mathematische Bildung: Mathematiklernen der Drei- bis Achtjährigen* [Early mathematics education: Mathematical learning of three- to eight-year-old children]. Berlin: Springer Spektrum. https://doi.org/10.1007/978-3-8274-2633-8

Bishop, A. J. (1988a). Mathematical education in its cultural context. *Educational Studies in Mathematics, 19*, 179–191. Retrieved from www.jstor.org/stable/348257

Bishop, A. J. (1988b). *Mathematical enculturation: a cultural perspective on mathematics education.* Dordrecht: Kluwer.

Björklund C., Magnusson M., & Palmér H. (2019). Teachers' involvement in children's mathematizing – beyond dichotomization between play and teaching, *European Early Childhood Education Research Journal, 26*(4), 469–480. https://doi.org/10.1080/1350293X.2018.1487162

Blömeke, S., Dunekacke, S., & Jenßen, L. (2017). Cognitive, educational and psychological determinants of prospective preschool teachers' beliefs. *European Early Childhood Education Research Journal, 25*(4), 1–19. https://doi.org/10.1080/1350293X.2017.1380885

Blömeke, S., Gustafsson, J.-E., & Shavelson, R. J. (2015). Beyond dichotomies. *Zeitschrift für Psychologie, 223*(1), 3–13. https://doi.org/10.1027/2151-2604/a000194

Boaler J., Chen L., Williams C., & Cordero M. (2016). Seeing as understanding: The importance of visual mathematics for our brain and learning. *Journal of Applied & Computational Mathematics, 5*(5), https://doi.org/10.4172/2168–9679.1000325. Retrieved from www.omicsonline.org/open-access/seeing-as-understanding-the-importance-of-visual-mathematics-for-our-brain-and-learning-2168-9679-1000325.pdf

Bruner, J. S. (1966). *Toward a theory of instruction.* Cambridge, Mass: Belknap Press.

Bruner, J. S. (1978). The role of dialogue in language acquisitions. In A. Sinclair, R. J. Jarvelle & W. J. M. Levelt (Eds.), *The child's concept of language.* New York: Springer-Verlag.

Bruns, J., Eichen, L., & Gasteiger, H. (2017). Mathematics-related competence of early childhood teachers visiting a continuous professional development course: An intervention study. *Mathematics Teacher Education and Development, 19*(3), 76–93.

Cheeseman, J. (2009). Challenging mathematical conversations. In R. Hunter, B. Bicknell, & T. Burgess (Eds.), *Crossing divides (Proceedings of the 32nd annual conference of the Mathematics Education Research Group of Australasia, Vol. 1),* (pp. 113–120). Palmerston North, NZ: MERGA. Retrieved from www2.merga.net.au/documents/Cheesman_RP09.pdf

Clements, D. H. & Sarama, J. (2009). *Learning and teaching early math: The learning trajectories approach.* New York: Routledge

Cooke, A., & Bruns, J. (2018). Early childhood educators' issues and perspectives in mathematics education. In E. Iliada, J. Mulligan, A. Anderson, A. Baccaglini-Frank, & C. Benz (Eds.), *Contemporary research and perspectives on early childhood mathematics education,* (pp. 267–290). ICME-13 Monographs: Springer International Publishing AG.

Dahl, R. (1970). *Fantastic Mr Fox.* London: George Allen & Unwin.

Devlin, K. J. (2012). *Introduction to mathematical thinking.* Palo Alto: Keith Devlin.

Dewey, J. (1934). *Art as experience.* New York: Putnam.

Diskowski, D. (2009). Bildungspläne für Kindertagesstätten — ein neues und noch unbegriffenes Steuerungsinstrument [Curricula for kindergartens – a new and still unapprehended control element.]. In H.-G. Roßbach & H.-P. Blossfeld (Eds.),

Zeitschrift für Erziehungswissenschaft Sonderheft: Vol. 11. *Frühpädagogische Förderung in Institutionen: Zeitschrift für Erziehungswissenschaft*, (pp. 47–61). Wiesbaden: VS Verlag für Sozialwissenschaften/GWV Fachverlage GmbH Wiesbaden. https://doi.org/10.1007/978-3-531-91452-7_4

Dockett, S., & Goff, W. (2013). Noticing young children's mathematical strengths and agency. In V. Steinle, L. Ball, & C. Bardini (Eds.). *Mathematics education: Yesterday, today and tomorrow (Proceedings of the 36th annual conference of the Mathematics Education Research Group of Australasia)*, (pp. 771–774). Melbourne, VIC: MERGA. Retrieved from https://files.eric.ed.gov/fulltext/ED572810.pdf

Dunekacke, S., Jenßen, L., Eilerts, K., & Blömeke, S. (2016). Epistemological beliefs of prospective preschool teachers and their relation to knowledge, perception, and planning abilities in the field of mathematics: A process model. *ZDM, 48*(1–2), 125–137. https://doi.org/10.1007/s11858-015-0711-6

Gasteiger, H. (2015). Early mathematics in play situations: continuity of learning. In B. Perry, A. MacDonald, & A. Gervasoni (Eds.), *Early mathematics learning and development: mathematics and transition to school: international perspectives*, (pp. 255–271). Singapore, Heidelberg, New York, Dordrecht, London: Springer.

Klibanoff, R. S., Levine S. C., Huttenlocher, J., Vasilyeva, M., & Hedges L. V. (2006). Preschool children's mathematical knowledge: the effect of teacher "math talk." *Developmental Psychology, 42*(1), 59–69. https://doi.org/10.1037/0012-1649.42.1.59

Kluczniok, K., Anders, Y., Sechtig, J., & Rossbach, H. -G. (2016). Influences of an academically oriented preschool curriculum on the development of children – are there negative consequences for the children's socio-emotional competencies? *Early Child Development and Care, 186*(1), 117–139. https://doi.org/10.1080/03004430.2014.924512

Lüken, M., Peter-Koop, A., & Kohlhoff, S. (2014). Influence of early repeating patterning ability on school mathematics learning. In P. Liljedahl, C. Nicol, Oesterle, Susan, & D. Allan (Eds.), *Proceedings of the 38th Conference of the International Group for the Psychology of Mathematics Education and the 36th Conference of the North American Chapter of the Psychology of Mathematics Education*, (pp. 137–144). Vancouver: PME.

McCray, J. S., & Chen, J.-Q. (2012). Pedagogical content knowledge for preschool mathematics: Construct validity of a new teacher interview. *Journal of Research in Childhood Education, 26*(3), 291–307. https://doi.org/10.1080/02568543.2012.685123

Ministry of Education and Research. (2012). *Forskrift om rammeplan for barnehagelærerutdanning* [Regulation of framework plan for preschool teacher education]. Oslo: Ministry of Education and Research. Retrieved from www.regjeringen.no/globalassets/upload/kd/rundskriv/2012/forskrift_rammeplan_barnehagelaererutdanning.pdf.

Ministry of Education and Research. (2017). *Framework plan for kindergartens – content and tasks.* Oslo: Ministry of Education and Research. Retrieved from www.udir.no/globalassets/filer/barnehage/rammeplan/framework-plan-for-kindergartens2-2017.pdf.

Nguyen, T., Watts, T. W., Duncan G. J., Clements D. H., Sarama, J. S., Wolfe, C., & Spitler, M. E. (2016). Which preschool mathematics competencies are most predictive of fifth grade achievement? *Early Childhood Research Quarterly, 36*, 550–560.

Oppermann, E., Anders, Y., & Hachfeld, A. (2016). The influence of preschool teachers' content knowledge and mathematical ability beliefs on their sensitivity to mathematics in children's play. *Teaching and Teacher Education, 58*, 174–184. https://doi.org/10.1016/j.tate.2016.05.004

Perry, B., & Dockett, S. (2005). What did you do in maths today? *Australian Journal of Early Childhood, 30*(3), 32.

Peter-Koop, A., & Scherer, P. (2012). Early childhood mathematics teaching and learning. *Journal für Mathematik-Didaktik, 33*(2), 175–179. https://doi.org/10.1007/s13138-012-0043-9

Piaget, J., & Inhelder, B. (1948). *La représentation de l'éspace chez l'enfant* [Childrens representation of space]. Paris: Presses Universitaires de France.

Schuler, S. (2013). *Mathematische Bildung im Kindergarten in formal offenen Situationen: Eine Untersuchung am Beispiel von Spielen zum Erwerb des Zahlbegriffs* [Mathematical education in kindergarten in formally open situations: A study using the example of play for the acquisition of the concept of numbers]. Münster: Waxmann.

Strohmer, J., & Mischo, C. (2015). Does early childhood teacher education foster professional competencies? Professional competencies of beginners and graduates in different education tracks in Germany. *Early Child Development and Care, 186*(1), 42–60. https://doi.org/10.1080/03004430.2014.985217

Thiel, O. (2010). Teachers' attitudes towards mathematics in early childhood education. *European Early Childhood Education Research Journal, 18*(1), 105–115. https://doi.org/10.1080/13502930903520090

Ulferts, H., & Anders, Y. (2016). *Effects of ECEC on academic outcomes in literacy and mathematics: Meta-analysis of European longitudinal studies.* Utrecht: CARE project, Utrecht University. Retrieved from CARE homepage http://ecec-care.org/fileadmin/careproject/Publications/reports/CARE_WP4_D4_2_Metaanalysis_public.pdf

Unterhauser, E., & Gasteiger, H. (2018). Verständnis des geometrischen Begriffs Viereck bei Kindern zwischen vier und sechs Jahren [The understanding of the geometrical term square in children between the ages of four and six]. *Frühe Bildung, 7*(3), 152–158. https://doi.org/10.1026/2191–9186/a000382

Wheatley, G. H. (1997). Reasoning with images in mathematical activity. In L. D. English (Ed.), *Mathematical reasoning – analogies, metaphors, and images,* (pp. 281–297). Mahwah, NJ, US: Lawrence Erlbaum.

8

GHOST STAIRS AND A GHOST TREE

Elena Severina, Andreas Lade, and Zoi Nikiforidou

Vignette

The vignette comes from a research project aimed at seeing what opportunities to explore mathematical ideas could naturally occur when children engage with photography in kindergarten. At the beginning of the project, children were placed in groups of four or five and asked to take pictures of what they found nice in the outdoor environment. Later they were presented with the idea of making a photobook. The children eagerly took photographs, discussed which photos they liked and why, designed their own pages of a photobook on a touch screen computer and, finally, reviewed a hard copy of the photobook.

The project was implemented in one private Norwegian kindergarten with a nature and outdoor profile. This vignette tells the story of two photographs, following four Norwegian five-year-olds from a pre-school group. The preservice kindergarten teachers, Kari and Andreas, are Norwegians with different dialects. The researcher, Elena, is Russian but speaks fluent Norwegian.

Selecting photographs

Peter, Olav, Maria, and Inga are sitting on a sofa and viewing photos on a touchscreen computer in the teacher's common room. When the photo of stone stairs (Figure 8.1) appears on the screen, Olav calls it "ghost stairs," and Peter says, "I took … I was walking downstairs while I took this picture." All the children think it is a scary photo, and girls do not want to include it in the book, but the boys do. Later, they make a compromise: "ghost stairs" should be included if a "hairy stone" is included as well.

FIGURE 8.1 Ghost stairs

A photo of a tree (Figure 8.2) is on the screen. Peter and Olav agree that this is an old tree, "because it is … it missed many bushes," explains Peter. Olav says, "OK. Now I will investigate if this is a tree," and he climbs on the backrest, "Yes … It is kind of dark." Peter climbs to Olav. Andreas tries to tilt the screen, "Is it still dark?" Inga also climbs on the backrest. Olav says, "Yes." Andreas tilts the screen, and Inga slides from the top of the backrest to the seat of the sofa. Peter says, "Does it look completely like a ghost tree?" While Olav slides down, Andreas says, "Yes, it looks almost like a ghost tree." Inga and Maria try to get Peter down onto the sofa while Andreas asks if the children think that it is a cool picture. Olav says, "I think it is cool! Yes." The girls still struggle with getting Peter down. Andreas says, at the moment Peter finally slides down, "How is it cool?" Peter has an answer, "Because it can attack us." Andreas asks the children to sit down and decide if they want to use the picture in the photobook. Everyone wants to.

FIGURE 8.2 A ghost tree

Designing pages

Later, in the same room, Olav has been designing his layout on a touchscreen computer for a while. He moved, scaled, and rotated photographs and placeholders, and moved and scaled the whole book. The left image he placed upside down, the right on the top straight (Figure 8.3, left). He moves "the ghost tree" in the frame of the left image, releases it (Figure 8.3, centre), and the frame with "the ghost tree" rotates back to the straight position (Figure 8.3, right). Olav screams, "No!" and shakes his head. He touches the screen and moves the photobook. When it is back to the centre, Olav says, "Yes!"

Olav struggles a bit with the interface but finally manages to rotate "the ghost tree" upside down, "Like this" (Figure 8.4). Reminded by Andreas, Olav asks the others if he should include the "ghost stairs" as the third picture. The other children agree.

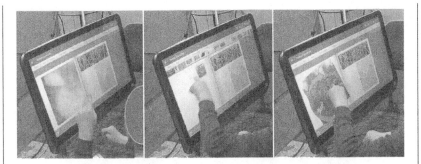

FIGURE 8.3 Designing the photobook

FIGURE 8.4 Olav's page

Review photobook

It is a month later in the same room. Maria and Inga study the printed photobook together with Kari and Elena. Elena points at the photo on the left page and asks, "Do you remember this tree?" Maria says it is called "Old-old tree," and adds, pointing at the photograph, "It is upside down." Elena agrees. "Is this a tree?" wonders Inga, turning the photobook upside down (Figure 8.5). Inga concludes, "This is a tree," and Maria takes the photobook from the table (Figure 8.6) and moves it closer to her face. Inga and Kari also move in order to see the pictures.

Keeping the photobook upside down, Maria points at the "Dragon forest," saying to Kari, "Now they are upside down!" Kari says, "Hmm … like this."

FIGURE 8.5 Inga rotates the photobook clockwise with one hand until it is upside down

FIGURE 8.6 Reconstruction of Maria taking the book from the table.

Maria proceeds, pointing at the picture, "And then you see ... this round ... It looks like we are walking down..." (Figure 8.7, right). Maria turns the photobook counterclockwise to the straight position and continues, pointing at the picture (Figure 8.7, left), "And these here look like ... we are walking ... up." Inga follows the slope of the stairs with her finger in the air close to the page. Kari says, "And here we are walking up, yes." Maria turns the photobook towards the teacher, and Inga cannot see the pictures anymore. Inga holds the photobook at the top, Maria at the bottom. Inga's head is between the photobook and Maria. Kari asks, while Inga pushes the top of the photobook down so that it is placed almost horizontally (Figure 8.8), "So, if you turn the book ... you are walking ...?" Maria looks at Kari with a smile and says, "Down!" Both laugh.

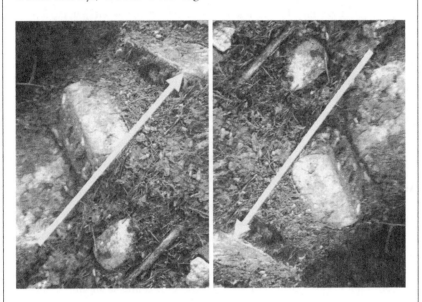

FIGURE 8.7 Reconstruction of Maria's pointing movement
Left: The photobook is straight. Right: the photobook is upside down

FIGURE 8.8 Reconstruction of Inga placing the photobook almost horizontally

Inga speaks with a low voice, "I will see. I will see. Hmm." Maria turns to Inga, and the girls hold the book together. Inga's face is close to the photograph. Maria says, "See, we are walking down...," and moves her finger diagonally downwards. Then Inga draws an "upstairs line" with her nose. The girls laugh. They look at the photobook a bit longer, talking indistinctly. Maria continues to move and rotate the photobook while she follows the stairs with her finger. Inga observes. A moment later the photobook is returned to the table. They continue to turn the pages.

Reflective questions

1   A range of specific physical actions accompany the children's interactions with these photos. In what ways might the children's embodied actions contribute to their mathematical understandings?
2   In what ways do these experiences with the photographs and the photobook contribute to the children's understanding of their spatial environment?
3   Why might the children seek to verify the image of the tree (on the screen) when they were involved in taking the image?

Commentary 1

Author: Elena Severina

The basis for the pedagogical practice in early childhood education in Norway is defined by the *Framework Plan for Kindergartens – Content and Tasks* (Ministry of Education and Research, 2017). According to it, all kindergartens must take a holistic approach to the child's development, where care, play, and learning are the core values, and both indoor and outdoor environments are important pedagogical arenas. Further, children's curiosity, creativity, and will to learn are seen as essential in forming learning processes, and therefore should be acknowledged and stimulated. The children's participation, their own play, and daily activities are highlighted throughout the Framework Plan, focusing on play-based learning instead of school-like teaching, following the socio-pedagogical tradition.

Expectations for the work of kindergartens within different subjects are described in the seven learning areas. One of them, "Quantities, spaces and shapes," is dedicated to mathematics. In the Framework Plan, mathematics is seen as a tool to "understand relationships in nature, society and the universe" (Ministry of Education and Research, 2017, p. 53). Kindergarten teachers are expected to create varied rich environments in order to stimulate children's sense of wonder, curiosity, and motivation for problem-solving and to enable the children to explore and discover mathematics in the world around them. Mathematical topics cover "play and investigation involving comparison, sorting, placement, orientation, visualisation, shapes, patterns, numbers, counting and measuring" (Ministry of Education and Research, 2017, p. 53). Asking questions, using reasoning, using argumentation, and seeking solutions are also associated with this learning area. The fundamental mathematical activities (Bishop, 1988) are not explicitly mentioned in the Framework Plan (Ministry of Education and Research, 2017, p. 53), but are commonly used to identify mathematics in children's activities in Norwegian kindergartens.

In the vignette, the children freely explore physical and virtual spaces, develop, share, and argue for their ideas, as well as pose and solve problems. The research project has a child-centred design in a technology-rich environment: photo camera, iPad, and photobook editing software on a touch screen computer are actively and willingly used by the children to develop and express their ideas.

Explaining is one of the fundamental mathematical activities (Bishop 1988) that children were expected to engage with by the project design. When asked to figure out something in small groups, they needed to explain their thoughts and bring arguments to support them. One of the phases of the project, "Selecting photographs," builds on classification as the children are asked to point out the images they want to include in the photobook, and explain why. For example, Andreas follows Olav's statement that the old tree photo is cool by asking, "How is it cool?" and Peter answers, "Because it can attack us"; or Peter explains why they think the tree is old by saying, "because it is … because it missed many bushes."

In the vignette, the "ghost tree" image (Figure 8.2) is dark, the paper in the photobook is glossy, and the touch screen computer has a rather limited view angle. I find it interesting that even though the image has already been recognised in a group, two children want to confirm that the object in the photo is actually a tree. First, during the photo selection, Peter and Olav seem to agree that this is an old tree in Figure 8.2 and even provide an explanation why it is old. Still, Olav says, "I will investigate if this is a tree" and climbs up on the backrest, changing his view angle towards the screen, and then concludes, "Yes … it is kind of dark." The next episode is during the photobook reviewing. Elena asks the children, "Do you remember this tree?" and Maria responds immediately, "It is called old-old tree" and comments that it is upside down. However, Inga wonders "Is this a tree?" turns the book clockwise 180 degrees (Figure 8.5), and concludes, "Yes, this is a tree." Here, both Olav and Inga spontaneously pose and solve a real problem of classification with the help of their spatial skills: being challenged by the view angle and darkness of the object (the tree) in the two-dimensional space (the photo), they adjust their position in the three-dimensional space (the room). The whole process is completely driven by the children and is supported by oral utterances and gestures. Schoolchildren often stop caring about a problem after knowing the right answer. Here, Olav and Inga seem to know that it is a tree in the photograph, and they still care about verifying it as they have their concerns. In these episodes, the fundamental activity of explaining is intertwined with locating (Bishop, 1988).

Locating activity is described by Bishop (1988) as "exploring one's spatial environment and conceptualising and symbolising that environment, with models, diagrams, drawings, words or other means" (p. 182). According to Bishop (1988), spatial orientation is a part of locating that involves understanding and operating with relationships between different positions in space; it develops first with respect to one's own position and one's own movement through space (e.g. Olav's and Inga's space actions described above), and progresses towards being able to use maps and coordinates (Sarama & Clements, 2009). Spatial visualisation is a part of locating that includes creating and manipulating mental images of two- and three-dimensional objects, as well as using these images in order to think, communicate, and develop new ideas (Bishop, 1991).

According to Berthelot and Salin (1998), space can be seen as having three representations: macrospace, mesospace, and microspace, and experiences with all three of them are important in order to develop spatial awareness and the ability to use school geometry in real life. Føsker (2012, p. 75) exemplifies these spaces in the kindergarten context: a macrospace can be an unknown forest or a city; a mesospace can be a room in the kindergarten building or outdoor area; and a microspace can be a sheet of paper or a table. The vignette has examples of how children can engage with the concept of rotation and such spatial concepts as "down," "up," and "upside down" in micro- and mesospaces, both virtual and physical.

When Olav works on the touch screen with the photobook design, he operates in a microspace of the editing tool. He has received minimal instructions and is exploring this virtual space while moving, scaling, and rotating photographs and placeholders, and moving and scaling the whole book. Olav spends some time exploring the programme, smiling and commenting on his actions, but then he seems to have got a plan. Judging from his oral utterances and dedication to the process, as well as manipulations on the touchscreen, Olav seems to have a clear idea about the orientation of the photograph (he screams "No!" and shakes his head when the tree rotates to the normal position, see Figure 8.3). He rotates the placeholder with the photograph in both directions with slightly different gestures each time until he manages to place the image upside down, probably matching his mental prototype ("Like this!"). The possibilities of the software were both supportive and challenging throughout this process. Here Olav solves the problem of how to turn the image upside down. Although the placing of the image according to the mental prototype can be classified as designing (Bishop, 1991), the experience Olav gains with rotation, scaling, and movement of the objects in the virtual space belongs to the domain of locating (Bishop, 1988). Here the connection between the terms "upside down" and "rotation" is established through experience with rotation as a spatial visualisation and as a bodily action on an object in a virtual space.

Another exploration of the relationship between rotation and the upside-down position of the photo can be found during the photobook review. After Inga turns the photobook upside down, Maria moves it very close to her face. This gives her a new perspective (she kind of zooms into the microspace of the photobook) and she notices that now two other pictures are upside down. Maria's excitement ("Now they are upside down!") could indicate that it is unexpected and possibly challenging her spatial visualisation skills. Earlier, when the children worked on the computer with page design, they could flip and rotate the pictures independently. Now, after the printed photobook was rotated to place the "ghost tree" photo straight, two other images are turned upside down. According to variation theory (Marton, Runesson, & Tsui, 2004), in order to learn about a certain aspect of a phenomenon, it is necessary to experience variation in the dimensions that correspond to that aspect. Here the difference in the behaviour of the virtual space of the editing tool and the printed photobook seems to create a dimension of variation, where the relationship between rotation and the "upside-down" position becomes visible.

Further, the situation develops into a playful exploration of the effect that rotation of the printed photobook has on the orientation of the objects in the photographs. Maria proceeds to study the microspace of the photobook, this time focusing on the "ghost stairs." She explains to Kari that, when the photobook is upside down, "it looks like we are walking down," and follows the slope of the stairs with her finger (Figure 8.7, right). Then she turns the photobook back to the straight position and explains that now "these look like ... we are walking ... up," and follows the slope of the stairs with her finger (Figure 8.7, left). In both cases, the gesture and oral language are synchronised and communicate first "walk downstairs" and then "walk upstairs." Kari supports Maria's investigation by challenging her to explain again, and the conversation proceeds. Inga follows them with interest and now adjusts the position of the photobook (Figure 8.8) to see the image. Maria invites Inga to join her play by turning to her and repeating the explanation, and they proceed to experiment with rotation of the photobook, giggling for some time.

It is not completely clear why the girls associate the normal position of the photo with "walking upstairs," and the upside-down position of the photo with "walking downstairs." One possible explanation could be that Maria uses the white stone and/or the stick near the stairs to define the direction, using them like an arrow. Alternatively, it could be that in the photo the bottom part of the stairs is on the left, and the top part of the stairs is on the right, and this might suggest walking up (the same way as traffic sign indicating an "uphill road" suggests uphill). The girls could possibly associate the upside-down position of the photo with the "inverse" of walking up, that is, walking down. In both cases, interpretation is related to her previous experience with the symbolic language used in the culture.

In the situations discussed here, the children gained experience with three tangible spaces of different scale: mesospace of the room (real), microspace of the photobook (real), and microspace of the editing tool (virtual). Object behaviour and allowed manipulations/operations in these spaces vary. The movement of objects on the touch screen is fairly intuitive and rather similar to the real-life experience, while scaling is done more easily on a screen than in the physical world. Exploration of scaling may be used to talk about ratio and perspective. Rotation in different spaces depends on the object: one may easily rotate a toy car in one's hand or a photograph with an editing tool, but to rotate a tree upside down, one needs to stand on one's head, making the whole world appear upside down. Additionally, when it comes to editing tools, both the shape and the dynamics of gestures, selected object, and possibilities of combining rotation with scaling will affect the result.

During the situations described in the vignette, differences between the spaces helped create new dimensions of possible variation and made children curious and active in problem-solving. Experiences with different representations of the tree and the stairs in these spaces resulted in interesting mathematical conversations, which may have helped the children gain a deeper and more general understanding of concepts like rotation, scaling, right/left, and up/down.

Commentary 2

Author: Andreas Lade

The kindergarten in which the vignette originated has a focus on nature and out-door activities with an emphasis on experiences on trips beyond the kindergarten. This gave the project an enormous advantage as the children were already used to traversing nature and its rough terrain. Since the project's main event was picture taking in a botanical garden called "Arboretet," with many side roads and paths, the actual practical execution went flawlessly. Elena gave the kindergarten's educational supervisor the task of informing the children of the photobook project and its photo activities. As we arrived at the kindergarten and were set to make our first trip to explore nature, we gave the children a quick tutorial on photo etiquette, including how to care for the photographic tools that we had at our disposal, and that the children were not allowed to take photographs of one another's faces. The groups were given small tips and tricks on how to operate the photographic tools in an individual manner while on the trip.

While using photography in nature is a creative way to get children interacting, discussing, and exploring the different photography devices with each other, it made me aware of my position in this project. When I am defining my roles as a prospective kindergarten teacher in the project, there are two definite sides to the matter: the passive project participant and the active kindergarten teacher. When performing the activities connected to the project, I had to slip carefully between a passive role and an active role. The passive project participant's role was not just to observe the children's technical skills with the photography tools but also their mathematical skills. This involved active listening and observing their thoughts and discussions in their democratic decision-making. The active kindergarten teacher's role was to support the children in their traversing of the forest, to care for their well-being, and to support them in their activities as kindergarten staff.

It proved itself strange to manoeuvre those conflicting roles when, as a future kindergarten teacher, you need to adhere to the wording in the *Framework Plan for Kindergartens – Content and Tasks* (Ministry of Education and Research, 2017, p. 19), "Kindergartens shall actively encourage caring relationships between children and staff and between the children themselves in order to foster well-being, happiness and achievements," and, as an observing and listening project participant, you should not interfere with the children's activities. Both roles are needed in order to success-fully achieve the desired result. They should merge together as scaffolding (Wood,

Bruner, & Ross, 1976). Both teach in the moment and provide support when the children need it. The picture in Figure 8.1 would not have come to pass if the children's discussions had not been directed to the staff. While Olav and Peter were in control of the photography device, they were undecided on whose turn it was to operate it. As they argued, and could not seem to find a solution, they asked the adults for guidance. The roles blurred, and I took on both roles, guiding them while supporting their own decision. This ended in Peter taking the picture.

As the children were not familiar with all the techniques involved in using the equipment, this might have benefited the project in an unexpected manner. While using the tablet for photographing the upside-down tree (Figures 8.2, 8.3 and 8.4), the process of taking the photo led to a rapid learning curve about how to operate the tablet. Olav observed the tree from a distance and went closer to take the picture without using his fingers to zoom. As this is fascinating in itself, he also tried to take a picture upside down. The tablet did not correct and flip the angle to the normal level as it does on the phone. After first taking the photograph of the tree normally, Olav studied it for a while. He then flipped the tablet and inspected it in view mode. He let out a little giggle and went back to photo mode. He then took a new picture while the tablet was upside down. This resulted in the picture in Figure 8.2 being intentionally taken upside down to mimic the skull of a human being. Now this might be a happy coincidence or a stroke of genius. He noticed right away in the picture selecting process that he could rotate the pictures at will, but his skull picture had to be upside down. Olav was showing clear signs of spatial reasoning. He had a set idea of what he wanted out of the picture. He investigated and oriented himself in the picture, and made a mental transformation that formed it deliberately to his liking, thus creating something entirely different out of his imagination. The tree became a skull out of his understanding and imagination.

Figure 8.2, a "ghost tree" in the vignette, was intensely discussed by the children. It almost had a fairy-tale-like way of enthralling the children, especially the boys. When Olav, Peter, Maria, and Inga examined the picture in the photo selecting process, they explored the idea of spatial awareness. They all seemed to investigate the photo from both close and afar to understand fully what was going on.

The children drew their assumptions about the age of the tree from its lack of branches and its darker colour palette. As they drew conclusions on the tree, we can see that they undertook activities that are mentioned in the Framework Plan, "Kindergartens shall highlight relationships and enable the children to explore and discover mathematics in everyday life, technology, nature, art and culture and by being creative and imaginative" (Ministry of Education and Research, 2017, p. 53). This sentence encompasses what the children were doing in this process – exploring mathematics in everyday life, in both free play and in these "laboratory" activities. By *laboratory*, I mean the closed-off, small-group activities that have a narrow focus on an achievement or goal. The children work in these small groups using their imagination and creativity to explore mathematics through technology in their democratic discussions. It ended up becoming the "ghost tree."

In Figure 8.3, a month later, the same picture popped up in the photobook editing process, when the children were carefully designing their pages. In a frustrated attempt to place the photograph that Olav took of the "ghost tree" into the photobook, he now had to undo his clever use of upside-down photographing. As the picture appeared the "wrong way," he had to manoeuvre his fingers to adjust the angle of the picture to his liking, which was the original upside-down positioning. He had to find the right placement. Centring of the picture proved to be his biggest challenge as his hand-eye coordination was tested. As we observe, we can quickly recognise Bishop's (1988) locating through his execution and understanding. The locating takes form in the use of his fingers to manoeuvre the picture following his plan to place the picture in a set position. He also exhibited the vision to place himself into the picture to give it a new meaning (the "ghost tree").

In Figures 8.5, 8.6, and 8.7. there is yet another time lapse, and now the girls, Inga and Maria, are examining the finished product that is the photobook. They seem to be baffled by the visual effects of the ghost stairs. It is somewhat of an optical illusion, but they seem to be highly certain of the right way – this is up, and that is down. Now as Inga follows her finger down the line of Figure 8.7 (right), you can clearly see that she is drawing a diagonal line along the steps. As Maria turns the book, she now looks at it from an opposite point of view to Inga and draws a line up the stairs (Figure 8.7, left). Maria starts in one corner of the picture, and Inga starts in the other.

They "enter" the picture to view themselves from their own perspective, displaying spatial reasoning and locating skills proposed by Bishop (1988). They are fully aware of the picture's duality and play around with it, exploring the picture.

My final thoughts as a whole are that the project was very beneficial to the children, even though it had a slow start: allowing them to explore and further their curiosity in their environment; using modern photography tools in their quest for the perfect picture; allowing them to express their thoughts and beliefs through dialogues in their collective group; making decisions together; and challenging each other in their picture selections. Throughout the project, one could notice their actions and talking points were rooted in what Bishop (1988) calls "locating." They considered how they perceived the images and how they saw themselves inside the picture. As locating can be explained as "viewing oneself from *within* the space" (Sando, 2017, p. 40, italics in original, the author's translation), it also applies to pictures. The children were observed imagining themselves within the photographs while taking or talking about them. One example can be extrapolated from Figure 8.7 in the vignette where the stairs are visualised as going both upwards and downwards. Both Inga and Maria are visualising themselves within the picture and interpreting their view of the picture's reality. Their reality of the picture depends on both the angle and the rotation of the book. They are figuratively looking at and examining the picture to determine their location within the picture.

The next step could be an introduction to coding for children; a simplified visual programming tool with pictures that the children take on their own; pictures of ongoing activity from start to finish, where they need to arrange pictures in a set order to make a coherent sequence; to observe the whole activity from an outside perspective (taking pictures), and later integrate themselves into the actual sequence making process. This is an evolved form of the "viewing oneself from *within* the space" comment made earlier. They now have extensive experience taking pictures on the children's level, so it might be a good stepping stone. Perhaps they could make a topographical map of their neighbourhood using real pictures of their houses or points of interest. This allows the children to immerse themselves in their surroundings to create a mental map. This could evolve into treasure maps, where they often replace their pictures with symbols of the "x" and dotted lines that mark the way. Fictive maps, as much as real maps, help develop children's mental mapping, spatial reasoning, and spatial awareness. The possibilities are endless if one as a pedagogue is willing to explore and think outside the box. We are the shepherds of children's intellectual beginnings, so we have the privilege of challenging and putting their imagination to the test. They will construct an enormous amount of knowledge if we as kindergarten teachers dare to explore with them.

Commentary 3

Author: Zoi Nikiforidou

This vignette highlights how children are driven by their own ideas, interests, and artefacts (the photos they took) to explore and negotiate their thoughts and underpinning mathematical notions. Children were immersed in the activity and articulated their own views, even if they were conflicting, through verbal and embodied ways. During the selection of photographs, the design of the online photobook, and the review of the hardcopy photobook, they shared their observations and explanations before reaching decisions. This vignette illustrates outstanding pedagogical practice, including a variety of resources, a rich and stimulating environment, group work, child-oriented tasks, and opportunities that encourage children to talk about and share their mathematical ideas and strategies in a responsive context (Vogt et al., 2019). These learning experiences could have taken place in a Greek early childhood setting as, according to the *Single Cross Thematic Curriculum Framework* for pre-primary education and the relevant Analytical Curriculum (Ministry of Education, Research and Religious Affairs, 2003),

> programmes are designed around themes that trigger the children's interest while using pedagogic practices. Emphasis is placed upon an interdisciplinary and holistic approach to knowledge as well as making use of children's curiosity and ideas for conducting the learning process.
>
> *(Author's translation)*

The vignette shows how co-construction of knowledge and understanding (Vygotsky, 1962) happens, where kindergarten children made sense of the task and used the means of the camera and photography to create a photobook through interactions with each other. This approach of social constructivism and mathematics enculturation is prevalent in many countries, and as Bruner (1986) puts it,

> Most learning in most settings is a communal activity, a sharing of the culture. It is not just that the child must make his knowledge his own, but that he must make it his own in a community of those who share his sense of belonging to a culture.
>
> *(p. 86)*

In this vignette children were part of the shared culture and context in which the photographs were taken and analysed. Together, they developed a deeper level of reasoning and explanation by collectively exchanging opinions and perceptions. Furthermore, the children felt secure and comfortable to express their views. This indicates how this sense of belonging is important in enabling contribution and participation in dialogue with peers.

The children in this vignette experienced mathematics as part of their understanding of the world and thus, mathematics is viewed as "a way of knowing" (Bishop, 1991, p. 1). The prevalent mathematical activity that was witnessed is that of locating, "exploring one's spatial environment and conceptualising and symbolising that environment, with models, diagrams, drawings, words or other means" (Bishop, 1988, p.182). Initially, in the selection of the first photo (Figure 8.1), the decision and "explaining" process is based on what Andreas remembers and shares. The second photo (Figure 8.2) gives space for more negotiation, and participants get involved through their bodily and physical movements before deciding whether to include the "ghost tree" or not.

Through the online design of the photobook, the children moved, rotated, scaled, and created a "mental template" and overall engaged with the "designing" mathematical activity proposed by Bishop (1988). We can see how Olav, with his fingers, creates and recreates the layout of his page. Similarly, a month later, we see Maria, Inga, Elena, and Kari engaging with the hardcopy photobook and getting actively involved in occurring spatial concepts. They turn the book upside down, they point, they talk, and through enjoying themselves they explore spatial representations. The children in this activity (Figures 8.5–8.8) organically engaged with the photobook at a later stage based on their previous experiences and the photobook context that had a meaning for them. This follow-up activity underlines two key pedagogical principles. First, how mathematics and spatial reasoning can be enhanced in occurrences where children are stimulated and feel connected to the real world; in this case to their own or their peers' artefacts and photographs. Second, how manipulatives can enrich learning where senses, action, and perception get combined (Nikiforidou, 2019).

Mathematics manipulatives, either concrete or digital, are handled in "a sensory manner during which conscious and unconscious mathematical thinking is fostered" (Swan & Marshall, 2010, p. 14). This was evident in the vignette where children used the camera to produce digital and hardcopy pictures that provoked discussions and explorations in different directions. Both concrete and digital photos provided children with the opportunity to expand their thinking, and in both cases, the discussions might not have been similar, but were equally valid and valuable. This example reinforces the argument that mathematics manipulatives should not lie in the dichotomy of either concrete or digital, but instead in the consideration of both and how they can be effectively incorporated in meaningful activities and daily practice (Laski et al., 2015; Nikiforidou, 2019). Furthermore, manipulatives per se do not carry the meaning of the mathematical idea explored. It is the active, sensorimotor engagement of the learners within the wider pedagogical context that carries this meaning as evidenced by the participants in the vignette.

The role of the adult in supporting children's mathematical experiences in the classroom is crucial. As highlighted in Commentary 2, the adult can serve as a "scaffolder" (Wood, Bruner, & Ross, 1976) who simultaneously can provide support and guidance. In this direction, "guided play" (Fisher et al., 2013; Weisberg, Hirsh-Pasek, & Golinkoff, 2013), which sits between free play and direct instruction, is proposed as an effective educational approach to mathematics in early childhood. In guided play, adults structure the environment with manipulatives, materials, and other ingredients, but the children have control of the activities. Guided play combines the exploratory nature of free play, it is child directed and incorporates adult scaffolding. In this vignette, children had ownership of their learning, and adults had set the environment with materials, stimuli, and provocations. The children decided which photographs to use and why, but the adults provided them with the cameras, tablets, hardcopies, and the broader task to take outdoor pictures of something they found nice. This approach could be seen in a Greek setting, too, as the teacher's main purpose "is to support [children's] efforts by becoming an assistant, a colleague or a mediator" (Ministry of Education, Research and Religious Affairs, 2003, p. 254). The teacher, according to the Greek guidelines, creates the appropriate conditions for ensuring learning incentives and prerequisites for all children in an attractive, safe, friendly, and stimuli-rich environment.

This activity could be extended to further symbolic and creative activities. This would be supported by the Greek preschool cross-thematic pedagogy where through a project approach, children would gain a holistic perception of knowledge through meaningful and purposeful activities related to language, mathematics, studies of the environment, creation and expression (fine arts, drama, music, physical education), and computer science. In this direction, the children could be provided with materials and loose parts and encouraged to represent the photographs in different ways. Precisely, the children could use materials like clay, cartons, ropes, and stones to design the two pictures as concrete microworlds. They could also represent the photographs through virtual microworlds, as proposed in Commentary 2, through programming or topological mapping. In addition, they could extend their thinking and fantasy by creating stories, figures, and narratives based on the "ghost stairs" and the "ghost tree." Besides storytelling, they could role-play the stories, considering linguistic and mathematical notions in relation to time, location, size, orientation, and sequence. In this direction, mathematics would be merged into play "the 'as if' of imagined and hypothetical behaviour" (Bishop, 1988, p. 24). Consequently, children could imagine something, which is the basis for thinking hypothetically and beginning to think abstractly; model, which means abstracting certain features from reality; formalise and ritualise rules, procedures, and criteria; predict, guess, estimate, and assume what could happen; and explore numbers, shapes, dimensions, positions, and arguments, which in turn would mean engaging in the other five mathematical activities in playful ways.

References

Berthelot, R., and Salin, M. H. (1998). The role of pupil's spatial knowledge in the elementary teaching of geometry. In C. Mammana and V. Villani (Eds.), *Perspectives on the teaching of geometry for the 21st century* (pp. 71–78). Dordrecht: Kluwer Academic Publishers.

Bishop, A. J. (1988). Mathematical education in its cultural context. *Educational Studies in Mathematics, 19*, 179–191. Retrieved from www.jstor.org/stable/348257

Bishop, A. J. (1991). *Mathematical enculturation: A cultural perspective on mathematics education.* Dordrecht Kluwer Academic Publishers.

Bruner, J. (1986). *Actual minds, possible worlds.* Cambridge, MA: Harvard University Press.

Fisher, K., Hirsh-Pasek, K., Newcombe, N., & Golinkoff, R. M. (2013). Taking shape: supporting preschoolers' acquisition of geometric knowledge through guided play. *Child Development, 84*(6): 1872–1878.

Føsker, L. I. R. (2012). Grip rommet! Barns utvikling av romforståelse og barnehagelærerens systematiske arbeid med det [Catch the space! Children's development of spatial understanding and kindergarten teachers' systematical work with it]. In T. Fosse (Ed.), *Rom for matematikk – I barnehagen* [Space for mathematics – in the kindergarten], (pp. 61–89). Bergen: Caspar Forlag.

Laski, E. V., Jor'dan, J. R., Daoust, C., & Murray, A. K. (2015). What makes mathematics manipulatives effective? Lessons from cognitive science and Montessori education. *SAGE Open, 5*(2), 1–8. https://doi.org/10.1177/2158244015589588.

Marton, F., Runesson, U., & Tsui, A. (2004). The space for learning. In F. Marton & A. Tsui (Eds.), *Classroom discourse and the space for learning* (pp. 3–40). Mahwah, NJ: Erlbaum.

Ministry of Education and Research. (2017). *Framework plan for kindergartens - content and tasks.* Oslo: Ministry of Education and Research. Retrieved from www.udir.no/globalassets/filer/barnehage/rammeplan/framework-plan-for-kindergartens2-2017.pdf.

Ministry of Education, Research and Religious Affairs. (2003). *Single Cross Thematic Curriculum Framework (DEPPS).* Ministry of Education (Greece).

Nikiforidou, Z. (2019). Digital manipulatives and mathematics. In C. Gray and I. Palaiologou (Eds.), *Early learning in the digital age,* (pp. 210–223). London: SAGE.

Sando, S. (2017) Fundamentale matematikkaktiviteter [Fundamental mathematical activities]. *Tangenten – tidsskrift for matematikkundervisning, 28*(4), 38–45.

Sarama, J. A., & Clements, D. H. (2009) *Early childhood mathematics education research: Learning trajectories for young children.* New York: Routledge.

Swan, P., & Marshall, L. (2010). Revisiting mathematics manipulative materials. *Australian Primary Mathematics Classroom, 15*(2), 13–19.

Vogt, F., Hauser, B., Stebler, R., Rechsteiner, K., & Urech, C. (2019). Learning through play – Pedagogy and learning outcomes in early childhood mathematics. *European Early Childhood Education Research Journal, 26*(4), 589–603. https://doi.org/10.1080/1350293X.2018.1487160

Vygotsky, L. S. (1962). *Thought and language.* Cambridge, MA: MIT Press.

Weisberg, D. S., Hirsh-Pasek, K., & Golinkoff, R. M. (2013). Guided play: Where curricular goals meet a playful pedagogy. *Mind, Brain, and Education,* 7(2): 104–112.

Wood, D. J., Bruner, J. S., & Ross, G. (1976). The role of tutoring in problem solving. *Journal of Child Psychiatry and Psychology,* 17(2), 89–100.

PART 5

Designing

9

BUILDING BRIDGES BETWEEN MATHS AND ARTS

Lucía Casal de la Fuente, Carol Gillanders,
Rosa María Vicente Álvarez, and Camilla N. Justnes

Vignette

The official curriculum for preschool education in Galicia (Spain) for children aged three to six includes three content areas: self-knowledge and personal autonomy; knowledge of the environment and languages: communication and representation. The regional regulations pursue a holistic approach and emphasise the close relationships between the building blocks. The following vignette details a session for a one-room rural school with six children aged between three and six years. The aim of the session was to link maths and music with the ultimate goal of promoting mathematical thinking in early childhood through music. In particular, we focused on geometric shapes.

It was a very special day. The teacher had told the children that Lucía would be coming to the school to pay them a visit. The children were really excited and kept looking at the door. At last, they heard the bell ring. Lucía came in and said, "Hello" to them and to their teacher. Then, she immediately sat on the floor and invited the children to sing with her. They all sat in a circle and Lucía taught them a song that she called "The Welcome Song." Afterwards, she introduced the activity "Circles in the Air" with a question, "Do you know what circles look like?" Most of the children answered, "No," so she showed them how to make a circle by making a circle in the air with the tip of her index finger. The children looked fascinated and reacted very quickly when she asked them to show her their preferred index finger. When she started to make another circle in the air, all the children copied this movement following the pace (Figure 9.1). But not Xián, the youngest one. Oh, no! Xián made many circles in the air, moving his arms very quickly. This little boy was only three years old and he had just started school two

---

FIGURE 9.1 Making circles in the air

months before. When Lucía highlighted Xián's knowledge about the form of the circles, one of the oldest, five-year-old, Eneko, started to do the same rapid movements that Xián had previously done, while Tomás, another five-year-old kept doing the movement of the circles in a quiet manner. Soon afterwards, Tomás and Eneko stopped moving their arms. Whilst Xián tried to explain why he was doing this so quickly, Lidia, a four-year-old, started to copy the movements Xián had made, smiling and looking for Lucía's approval. This was exactly what Ainhoa, another four-year-old girl, did.

Suddenly, they all stopped making circles in the air when Lucía explained to them that they were going to accompany these movements with their voices. This was the first time they had been invited to combine vocal sound and movement. But this vocal sound (a rolled "r") was already familiar to the children. Before Lucía had emitted this rolled consonant, the children had started to do it themselves (even Lidia, a four-year-old girl, who could hardly pronounce it). Thereafter, Lucía emitted a glissando using this rolled "r" from low to high, and then from high to low, a sound that all the children immediately copied to accompany her. Trying to find a familiar situation connected to this sound, Lucía told them to imitate the onomatopoeic sound that a motorcycle's engine would make when being turned on. Immediately, Xián said it was like the sound of racing cars, and executed a high-pitched rolled "r" sound. Lucía smiled and said how nice that sounded. This time, she did the glissando with the rolled "r" while she drew a circle in the air with her index finger. The movement generated was a continuum of both the arm and the hand, in connection with the emanating voice, so that while making circles through gestures, she let her voice come out. The children copied her at once! Xián did it rushing once again, and Lucía asked him to do it more slowly. Then she invited all the children to do it once more, introducing the concepts "low" and "high," so that they could relate the sounds and movements (already interiorised and practised enough) to the specific musical terms. As expected, they followed her very accurately. She then asked them to pay attention to the shape and imaginary line they were drawing in the air, asking them about the name of that figure, to which the children answered that it was a circle. This time, the aim was to match the movements with the pitch.

Little by little, when the arms of the participants went up, the glissando slipped towards high pitches, and vice versa: when the arms went down, the glissando they emitted with their voices gradually fell to low pitches.

Suddenly, Xián ran towards a bag that Lucía had brought with her. Although it was closed, the little boy picked up a small red car and started playing by himself, mumbling and singing softly. Lucía could hear the racing car sound again. She touched her pocket and smiled. Yes, that was enough for today. She would use the slide whistle another time.

Reflective questions

1 How do the physical enactments of mathematical concepts develop the children's understanding of circles and their properties?
2 What do you notice about how the children develop the activity with their own play?
3 In what ways might you take this activity forward?

Commentary 1

Authors: Lucía Casal de la Fuente and Carol Gillanders

The current curriculum for Galician preschool education is regulated by the Galician Decree 330/2009 (Xunta de Galicia, 2009). It includes prescriptive guidelines and recommendations for all schools such as aims, contents, and methodological recommendations. This Decree emphasises the need to offer child-centred learning activities, attractive and varied resources, and a stimulating learning environment to reach the learning outcomes. Furthermore, it identifies the contributions to the development of basic skills. As happens in the rest of Spain, the enrolment rate of children aged three to six in Galicia in preschool education is nearly 100% (UNICEF Comité Galicia, 2018), although preschool education is not compulsory.

Early childhood and primary education in Spain is organised into three age-related periods of development: zero to three, three to six, and six to twelve years of age. The first two belong to early childhood education, and the last one to primary education. The aforementioned Decree is the principal reference for teachers in their day-to-day practices. Teachers design and develop their educational programmes taking into account the guidelines outlined by the regional government.

The activities were designed according to the contents and the development of fine motor skills. First of all, the children were asked to move, drawing big circles in the air, and were given time to do these movements. Drawing in the air was chosen because space offers no limits, and the standing position allows them to explore shape with their whole body. We must bear in mind Kellogg's (1969) contributions related to basic shapes when she states that children first draw big circles before being able to draw small ones. Once the children had explored the movements with their arms and index fingers, the teacher invited them to make a sound with a rolled "r." Afterwards, the teacher introduced one more suggestion: they should try to emit with their voices glissandos from low to high and vice versa whilst doing the movements with their arms. After some attempts, children were able to match the movements of arms and index finger to pitch, so that while the arm went from bottom to top (drawing a circle in the air), the glissando vocal sound went from low to high, and vice versa. Children enjoyed the activity and even proposed doing it with other consonants and movements. Finally, the teacher dedicated some time to explaining and naming the shape they had been making. Including opportunities for experiential learning is essential for successful learning (Dewey, 1934; Gardner, 2000; Moon, 2004), and the vignette is an example of this.

This vignette deals as well with emotional education at preschool. It shows how children at this age seek teachers' approval when approaching new materials or actions. It also illustrates some moments where feedback is provided by the teacher. Pekrun (2006) states that "the cognitive capacity needed to generate achievement-related causal expectancies and attributions, and related value appraisals, are acquired in the preschool and early elementary years" (p. 328), and highlights the importance that feedback on success and failure has on achieving outcomes. In this vignette, the teacher is very positive as well as firm and can maintain an effective learning environment.

In the above vignette, the teacher combined movement and voice, encouraging the children to accompany with sound the circles they drew in the air with their index finger. Bishop's (1988) fundamental activities were addressed as follows:

- *Locating:* Children had to draw circles in the air, using their arms and index fingers, stretching them right up and in front of themselves (Figure 9.1). This exercise involved exploring the surrounding space with high body control. To start and end a circle at the same point requires concentration and having interiorised the shape.

- *Measuring:* All the children followed the pace set by the teacher, but one little boy drew the imagined circles very fast. Comparisons could be explored regarding how many small circles could be made during the time it took to draw a big circle.
- *Designing:* Children created shapes in the air. Drawing imagined circles involved gross motor control (arm movements) as well as fine motor skills (the use of index fingers to draw the circles). Patterns and other shapes could be explored.
- *Playing:* While drawing circles in the air, the children were invited to emit glissando low-to-high sounds with their voices using a rolled "r," a phoneme that does not harm the voice. Imitations of onomatopoeic sounds such as motorbikes or the engine of a car were explored. Children had to follow the guidelines (high and low pitches). After the activity, one child took a toy car and started driving it while singing softly.
- *Explaining:* The indications given by the teacher were illustrative, not only with words but also with body language. Some children needed to verbalise the procedure of the activity undertaken, repeating out loud the explanation offered by the teacher.

Counting is the first of the fundamental activities indicated by Bishop (1988). Although this activity is not represented in the vignette, it would be possible to add it to the exercise; for instance, counting the number of circles done with each arm.

As can be seen, different contents from the areas of mathematics and music, in connection with other cross-cutting themes such as emotional and health education, are present in the activities reported in this vignette. Our vignette is especially focused on designing. The activities described can be done with other curricular contents, such as the form of the square, the rectangle, or the triangle. These three figures have sides, with straight line segments, whilst the circle does not, offering other opportunities for practising voice, movement, and shape (there is a growing body of research that focuses on the benefits of exploring music to teach other subjects, including Forgeard et al., 2008; Hallam, 2010; Jeanneret & Degraffenreid, 2018; Omidire et al., 2016; Rauscher, 2009; Shilling, 2002). This feature is of special interest, especially for working with the voice and for the development of pitch control, which has an impact on ear training as well. For instance, in a horizontal line, vocal pitch must be maintained as the line does not go up or down.

Materials and resources are other aspects to be commented on. Notwithstanding that, in the vignette, the only instruments used as learning tools were the body and the voice. Other resources could be used to attain the learning goals. For instance, exploring with hula hoops or driving toy cars along the perimeters of different shapes, as a child showed us when we implemented a set of activities in the same school. The use of handkerchiefs while making circles in the air at the pace set by a piece of music (particularly with soft tempos) is another possibility. The use of foam rubber tubes or colourful straws to form rectangles, triangles, and squares on the floor is a very attractive way to reinforce these mathematical contents. For example, we could ask the children to walk while the verses of a song are being played, and when the chorus starts, we could invite them to make some figures on the floor with those elements.

Commentary 2

Author: Rosa María Vicente Álvarez

The Spanish curriculum for the second period of preschool education (three to six years old) is established by a Royal Decree (Real Decreto 1630/2006, see Spanish Government, 2006). Although Spain has a decentralised state structure and each autonomous community has its own contextualised curriculum, this decree is a reference for all the Spanish regions including Galicia. The curriculum is open and should be taken into account when designing activities that integrate the key contents for preschool.

There are three areas of knowledge in preschool education in Spain, a stage in which contents and activities must be connected and worked jointly. We can see this in the activities and games presented in the vignette. These areas are as follows:

- *Self-knowledge and personal autonomy*, where block 2 refers to game and movement.
- *Knowledge of the environment*, where block 1 is related to the physical environment (where the geometric shape is included in the section "Elements, relationships and measurement").
- *Languages: communication and representation*. In the activities shown, the contents are presented in the following terms: block 1. Verbal language (listen, talk, and discuss); block 3. Artistic language (where we can situate concepts like form, space, sounds, and the possibilities of the voice and of one's own body, contrasts as long-short, strong-soft, fast-slow, and so on); and block 4. Body language, that could be linked to body expression games.

There are only six children in this session as it is a rural area. In Galicia, one-room schools are common, as this is a region with a high percentage of rurality. This is very thought-provoking since they have the possibility to learn from the experiences of others (Smit & Engeli, 2015). This situation is also frequent in many rural areas worldwide (Hargreaves, Kvalsund, & Galton, 2009), although the government tendency is to close one-room schools and to merge groups, putting together 25 children of the same age with one teacher into larger schools (with children from three to 16 years old), in the next larger town. As a result the mixed-age group setting (from three to six years old) is lost.

Two activities are shown in this vignette: the first one describes a game to get to know a geometric shape (the circle), with movement and voice associated with it. The other one shows an onomatopoeic game to work on the rolled consonant "r." During these games, different concepts like high-low, fast-slow, amongst others, are practised. Central to the assessment of any situation with young children is an awareness of the different developmental stages of consciousness in children.

Moving on to the goal of the activities, to promote mathematical thinking in early childhood through music, it is important to highlight the idea described in this vignette of combining vocal sounds and movement to prepare the children for learning different concepts, which is a very innovative idea. In this case, it can be very attractive and motivating both for children and adults. Researchers suggest that mathematics in preschool should include patterns (Björklund, 2008) and geometry (Clements & Sarama, 2009; Ginsburg, Lee, & Boyd, 2008). Games and exploratory activities are developed through the state's recommendations related to methodologies which are implemented with the support of Bishop's (1988) six fundamental activities. I highlight only two:

- *Measuring.* Comparing the time that it takes to make a circle in the air: some children adopted the rhythm indicated by Lucía whilst others did variations, making circles twice or triple fast. As well, they accompanied the concrete movements with their own voices (low and high) following abstract drawings (circles in the air).
- *Playing.* This is central in the curriculum and it is reflected in the vignette through language; mathematical concepts like shape, space, and own body; and sounds and possibilities of the voice following the indications of Lucía. As a result of this, children integrated the activities that are shown through cooperative games, exchanging ideas about the experience and learning from each other and finally, exploring for themselves without interaction of the teacher.

These kind of activities and games clearly show that it is possible to work from the educational field in an open, integrated, inclusive, and experiential way as the Spanish curriculum suggests. In addition, it is interesting to emphasise the reaction of the teacher, who gave time to the children to improve their abilities with the modification of the ideas presented in the welcome song and others. All games are focused on sound, voice, and movement to introduce an idea or maths concept. They all start from a known idea or pattern for the children, which helps contextualise the proposed tasks. From a practical point of view, the activities described are easy to combine in different ways such as introducing maths and other contents through symbolic songs and movements. Moreover, in the narrative vignette, these kind of games combined with specialised exercises and pedagogical experience could help teachers detect a disorder and the children to overcome it with attractive activities. Further research must be undertaken related to *Playing* and "teachers [sic] understanding of the activity to enhancing [sic] playful learning in early childhood education" (Svensson, 2015, p. 2008).

An interesting aspect is the kind of games described in the vignette which are appropriate to explore because they introduce maths with music and movement. They are clear, well explained, and the timing to develop activities is adapted to the children. Once the content has been practised, as well as different ways of approaching it, Lucía explores the activity conscientiously. In the vignette it is possible to appreciate that the activities end when the children have practised enough, changing the focus to other things. As we can see, the activities are also designed to be done by children with different talents. Children who master basic skills related to music and movement are given opportunities to make other patterns and to solve more challenging problems (Clements & Sarama, 2009). In this case, the interest in movement and sound associated during the practice of gestures brings as a result the enhancement of mathematical learning (Garber & Goldin-Meadow, 2002).

Concerning the focus on practical issues and educational possibilities, as can be seen, the children feel happy and they are waiting for Lucía (the researcher) with their arms open. The preparatory exercises seem to be well accepted by the children. From a practitioner's standpoint, it is relevant to highlight that the teacher is probably a very good singer, but this fact could be seen as a difficulty for teachers who do not sing well. Teachers' initial education is very important, and vocal training should be included as a core subject during this period. The principal reason is that these kinds of lessons with music are often discarded because teachers may think they are not good singers. However, here the teacher created a simple and complete activity with voice. In accordance with this idea, Burton (2005) suggests music educators should focus on developing positive attitudes in preschool teachers and in creating a higher comfort level for learning music activities through actual experiences in the classroom, as we can see in this vignette (Jordan-DeCarbo & Galliford, 2019). It seems clear that the predisposition of the children could be motivated by this quality of Lucía, too. The children, in all cases, imitated different ways to do the activities because they were motivated and reinforced by the teacher, who stressed the action of the children, who were smiling, looking for approval, and so on. This situation contributed to learning because it is natural that children continue enjoying the game, adapting it to other contexts, forgetting what is happening around themselves, as happened in the final section of this vignette (Csikszentmihalyi, 1997).

To give some pedagogical advice, from the reflective praxis point of view, I recommend that teachers do activities chosen according to difficulties in the classroom, activities that can offer many possibilities to explore concepts. It is important that teachers practise the activities before they implement them in the classroom so as to have previous experience related to the possible difficulties or mistakes that can arise. We have to really grasp other ways to help children learn in an experiential manner using only those resources and materials we need at each moment: body, voice, and environment in a collaborative way. That is why having previous practical experience (Robertson, 1999) is necessary for teachers to really understand and reflect on their own practice.

Commentary 3

Author: Camilla N. Justnes

The Norwegian kindergarten practice is situated in a social pedagogical tradition where childhood has an intrinsic value. Kindergartens in Norway support children's development by intertwining care, play, and learning. Children's freedom to explore their surroundings, including those of mathematics, is highly regarded. As in Spain, preschool education in Norway is not compulsory, but all children over one year are entitled to a kindergarten place. Ninety per cent of all children between one and five years in Norway attend kindergarten (Norwegian Directorate for Education and Training, 2016).

The pedagogical practice of kindergartens is described in *Framework Plan for Kindergartens – Content and Tasks* (Ministry of Education and Research, 2017). The mathematics domain has its own chapter called "Quantities, Spaces and Shapes," with explicit expectations for what the institution and the staff should provide in terms of mathematical opportunities. Although not formally acknowledged, the description of the learning area and how to work with the learning area can be traced back to Bishop's six mathematical activities (Bishop, 1988).

The goal for the teacher in this vignette was to promote mathematical thinking in early childhood through music, and with a particular focus on geometrical shapes. There are objectives in the Norwegian Framework Plan that correspond with this goal. For example, "By engaging with quantities, spaces and shapes, kindergartens shall enable children to investigate and recognise the characteristics of different shapes and sort them in a variety of ways" and "to use their bodies and senses to develop spatial awareness," and "Staff shall use books, games, music, digital tools, natural materials, toys and equipment to inspire children's mathematical thinking" (Ministry of Education and Research, 2017, pp. 53–54).

Bishop describes two fundamental activities referring to spaces and shapes: locating and designing. Locating refers to direction, orientation, navigation, and location. Designing is about shapes, patterns, art, and architecture. According to literature used in early childhood education in Norway, Norwegian kindergartens should be concerned with promoting activities that allow children to analyse properties, examine how shapes fit together, and develop an understanding of patterns, rather than just naming prototypes of different shapes (Carlsen, Wathne, & Blomgren, 2017; Nakken & Thiel, 2019; Solem & Reikerås, 2017). This approach corresponds with research on children's recognition of geometrical shapes. Findings suggest that children encounter problems when only offered visual prototypes, and that educators should develop activities that will provide children with opportunities to explore the property attributes of the shapes being studied. Such activities should include atypical examples (such as shapes of different sizes, orientation, aspect ratio, and skewness) as well as typical examples (Aslan & Aktaş, 2010).

Children explore, recognise, and find similarities and differences with objects they encounter in their play and everyday life (Solem & Reikerås, 2017, p. 73).

The shape of an object is an important feature. It makes us recognise the object, and the shape makes it looks different from or like other objects. Recognising and distinguishing between different shapes will give an overview and create structure in the children's world (Solem & Reikerås, 2017, p. 78). Without common classification and language, we have to describe the object every time. There are many different criteria to classify shapes and figures by. Such criteria can be open or closed, straight or curved lines, the number of corners, the angles' degrees, and so on.

According to van Hiele (1986), children first identify prototypes of basic geometrical figures like triangle, square, and circle. These prototypes are then used to identify other similar shapes. For example, a shape is a circle because it looks like a biscuit or a hubcap. If a shape does not sufficiently resemble its prototype, the child may reject the classification. The idea that shapes can be classified and named by their properties, in contrast to what they look like, involves a huge cognitive leap (Lago, Stossel, & Fostnot, 2016, p. 7).

In this vignette, the children are offered to explore properties and attributes, and not visual prototypes. We read about Lucía who introduces the activity "Circles in the Air" with a question, "Do you know what circles look like?" With this question, she guides the children's attention towards the properties of a circle. Children are invited to use their body, arm, finger, and voice to explore the circle shape. By using their arm, the teacher supports the children in constructing the properties of the circle shape. Round shapes are closed with one continuous curved line. In Norwegian, we have the concept *runding*, which means a round shape. This shape is closed, round, and has no corners, but is not necessarily a circle. A circle is distinguished from other round shapes in that the distance from any point from the curved line to the centre is constant (radius). The children in the vignette are drawing a circle in the air. They envision a starting point, make the outline of the circle, and end the movement at approximately the same point, gaining experience that the circle is closed.

Children often meet two figures with the same shape and two figures with the same shape and size. The first is called similar, the latter congruent. A circle is a circle regardless of its orientation or overall size, and children need to experience and explore this (Clements & Sarama, 2009). The teacher in the vignette will be making bigger circles due to a longer arm, so the activity gives room for the children to gain experiences with similar shapes. Asking questions like "Is my shape the same as your shape?" "Does it matter which direction we draw the circle in the air?" "Is it still a circle if we draw it clockwise or counterclockwise?" and "Are there other ways to make this same shape?" will also support children's exploration of similarity and congruence.

When children focus on the enclosure and boundaries of the shape and distinguish between closed and not-closed shapes, polygons are often all drawn the same. If we ask children to draw a circle and a triangle in the air, the shapes could look alike if they do not take the angles into consideration. In this way, a triangle could also be considered to be a *runding* or a circle. With such a focus, they recognise the shape topologically and not by Euclidian properties (Piaget, 1967 in Lago et al.,

2016). The Euclidian properties include points, lines, and angles. We could assume that the arm movement in drawing a triangle would include stops or change of direction to represent the angles in the triangle. By exploring the shape of the circle with arm movement, the teacher supports experiences with the critical attribute that the circle has no angles. One of the children wants to make the circle at a fast pace, which is possible since they do not stop to make angles or to change direction. It would probably be more difficult to make a triangle with angles in such a quick manner. The activity described could be done with other shapes like the square, the rectangle, and the triangle. The use of voice and arm movements would then complement a focus on Euclidian properties, and support the children in distinguishing different shapes by their properties.

One strand of research on children's geometrical thinking and learning in early childhood focuses on the role of gestures for understanding a mathematical concept (Elia & Evangelou, 2014). According to Sabena (2008), a definition of gestures suitable for mathematics learning can include "All those movements of hands and arms that students and teachers perform during their mathematical activities and which are not a part of any other action (i.e. writing, using a tool, […])" (p. 21). With this in mind, we can look at the vignette in terms of how arm movements could support the children's learning about the concept "circle." The children observe and mimic the teacher, and later observe and copy other children's more rapid movements. They do not need a picture or a cut-out circle to experience and discuss the shape, as the arm movements can represent the mathematical idea. It is well documented that different representations of the same concept contribute to a deeper comprehension of the concept (Duval, 2006). Gestures can be considered as a representational tool for many mathematical ideas and can contribute to the process of converting abstract concepts in a visual and concrete form (Elia & Evangelou, 2014). The arm movement in this vignette does not have a communicative intention but has a role of objectifying geometrical notions and has the potential to complement speech. But sometimes, movements and gestures come instead of verbally expressing thinking and learning, and this could lead to a simplification where terms and details disappear (Elia, 2018). However, my insight is limited to what is referred to in the vignette, all assumptions about how language complements the body movements to promote mathematical thinking can only be seen as considerations.

References

Aslan, D., & Aktaş, Y. (2010). Children's classification of geometric shapes. *Journal of Çukurova University Institute of Social Sciences, 19*, 254–270. Retrieved from www.researchgate.net/publication/283296057_Children's_Classification_of_Geometric_Shapes

Bishop, A. J. (1988). Mathematics education in its cultural context. *Educational Studies in Mathematics, 19*, 179–191. Retrieved from www.jstor.org/stable/348257

Björklund, C. (2008). Toddlers' opportunities to learn mathematics. *International Journal of Early Childhood, 40*(1), 81–95.

Burton, S. L. (2005). Children's musical worlds: Considering the preparation of early childhood educators. *Early Childhood Connections, 11*(1), 23–28.

Carlsen, M., Wahtne, U., & Blomgren, G. (2017). *Matematikk for barnehagelærere* [Mathematics for Kindergarten teachers] (3rd ed.). Oslo: Cappelen Damm AS.

Clements, D. H., & Sarama, J. (2009). *Learning and teaching early math: The learning trajectories approach*. New York: Routledge.

Csikszentmihalyi, M. (1997). *Creativity. The psychology of discovery and invention.* New York: Harper Perennial.

Dewey, J. (1934). *Art as experience.* New York: Putnam.

Duval, R. (2006). A cognitive analysis of problems of comprehension in a learning of mathematics. *Educational Studies in Mathematics, 61,* 103–131.

Elia, I. (2018) Observing the use of gestures in young children's geometric thinking. In I. Elia, J. Mulligan, A. Anderson, A. Baccaglini-Frank, & C. Benz (Eds.), *Contemporary research and perspectives on early childhood mathematics education* (pp. 159–182). ICME-13 Monographs. Cham: Springer.

Elia, I., & Evangelou, K. (2014). Gesture in a kindergarten mathematics classroom. *European Early Childhood Education Research Journal, 22*(1), 45–66. https://doi.org/10.1080/1350293X.2013.865357

Forgeard, M., Winner, E., Norton, A., & Schlaug, G. (2008). Practicing a musical instrument in childhood is associated with enhanced verbal ability and nonverbal reasoning. *PloS ONE, 3*(10), e3566. https://doi.org/10.1371/journal.pone.0003566

Garber P., & Goldin-Meadow S. (2002). Gesture offers insight into problem-solving in children and adults. *Cognitive Science, 26,* 817–831. https://doi.org/10.1207/s15516709cog2606_5

Gardner, H. (2000). *The disciplined mind: What all students should understand.* New York: Simon & Schuster.

Ginsburg, H. P., Lee, J. S., & Boyd, J. S. (2008). Mathematics education for young children: What it is and how to promote it. *Social Policy Report, 22*(1), 3–23. Retrieved from https://files.eric.ed.gov/fulltext/ED521700.pdf

Hallam, S. (2010). The power of music: Its impact on the intellectual, social and personal development of children and young people. *International Journal of Music Education, 28*(3), 269–289.

Hargreaves, L., Kvalsund, R., & Galton, M. (2009). Reviews of research on rural schools and their communities in British and Nordic countries: Analytical perspectives and cultural meaning. *International Journal of Educational Research, 48*(2), 80–88. https://doi.org/10.1016/j.ijer.2009.02.001

Jeanneret, N., & Degraffenreid, G. M. (2018). Music education in the generalist classroom. In G. E. McPherson & G. F. Welch (Eds.), *Music learning and teaching in infancy, childhood, and adolescence* (pp.178–195). New York: Oxford University Press.

Jordan-DeCarbo, J., & Galliford, J. (2019). The effect of an age-appropriate music curriculum on motor and linguistic and nonlinguistic skills of children three to five years of age. In S. L. Burton & C. Crump Taggart, *Learning from young children: Research in early childhood music.* London: Rowman & Littlefield Education.

Kellogg, R. (1969). *Analyzing children's art.* California: National Press Books.

Lago, M., Stossel, S., & Fosnot, C.T. (2016). *Baby's wild adventure: Shapes and navigation.* Scotts Valley, CA: CreateSpace Independent Publishing Platform.

Moon, J. A. (2004). *A handbook of reflective and experiential learning. Theory and practice.* London: Routledge.

Ministry of Education and Research. (2017). *Framework plan for kindergartens – content and tasks.* Oslo: Norwegian Directorate for Education and Training. Retrieved from www.udir.no/globalassets/filer/barnehage/rammeplan/framework-plan-for-kindergartens2-2017.pdf

Nakken, A. H., & Thiel, O. (2019). *Matematikkens kjerne* [The core of mathematics], (2nd ed.). Bergen: Fagbokforlaget.

Norwegian Directorate for Education and Training. (2016). *The education mirror.* Oslo: Norwegian Directorate for Education and Training. Retrieved from http://utdanningsspeilet.udir.no/2016/wp-content/uploads/2016/10/Utdanningsspeilet_2016_en.pdf

Omidire, M. F., Ayob, S., Mampane, R. M., & Sefotho, M. M. (2016). Using structured movement educational activities to teach mathematics and language concepts to preschoolers. *South African Journal of Childhood Education, 8*(1), a513. https://doi.org/10.4102/sajce.v8i1.513

Pekrun, R. (2006). The control-value theory of achievement emotions: Assumptions, corollaries, and implications for educational research and practice. *Educational Psychology Review, 18*(4), 315–341. Retrieved from https://link.springer.com/article/10.1007/s10648-006-9029-9

Rauscher, F. H. (2009). The impact of music instruction on other skills. In S. Hallam, I. Cross, & M. Thaut (Eds.), *The Oxford handbook of music psychology* (pp. 244–252). Oxford: Oxford University Press.

Robertson, D. L. (1999). Professors' perspectives on their teaching: A new construct and developmental model. *Innovative Higher Education, 23*(4), 271–294. Retrieved from http://citeseerx.ist.psu.edu/viewdoc/download?doi=10.1.1.606.1203&rep=rep1&type=pdf

Sabena, C. (2008). On the semiotics of gestures. In L. Radford, G. Schubring, & F. Seeger (Eds.), *Semiotics in mathematics education: Epistemology, history, classroom, and culture* (pp. 19–38). Rotterdam: Sense Publishers.

Shilling, W. A. (2002). Mathematics, music, and movement: exploring concepts and connections. *Early Childhood Education Journal, 29*(3), 179–184.

Smit, R., & Engeli, E, (2015). An empirical model of mixed-age teaching. *International Journal of Educational Research, 74*, 136–145. https://doi.org/10.1016/j.ijer.2015.05.004

Solem, I. H., & Reikerås, E. K. L. (2017). *Det matematiske barnet* [The mathematical child] (3rd ed.). Bergen: Caspar Forlag.

Spanish Government (2006). *Real Decreto 1630/2006, de 29 de diciembre, por el que se establecen las enseñanzas mínimas del segundo ciclo de Educación infantile* [Royal Decree 1630/2006, 29th December, that establishes the curriculum for the second cycle of early childhood education]. Retrieved from www.boe.es/eli/es/rd/2006/12/29/1630

Svensson, C. (2015). Preschool teachers' understanding of playing as a mathematical activity. *CERME 9 - Ninth Congress of the European Society for Research in Mathematics Education,* Charles University in Prague, Faculty of Education. Czech Republic, 2003–2009. Retrieved from https://hal.archives-ouvertes.fr/hal-01288509/document

UNICEF Comité Galicia. (2018). *La infancia en Galicia 2018* [Childhood in Galicia 2018]. Santiago de Compostela: UNICEF Comité Galicia.

van Hiele, P. M. (1986). *Structure and insight. A theory of mathematics education.* Orlando: Academic Press.

Xunta de Galicia (2009). *Lexislación da Educación Infantil en Galicia* [Galician educational laws for preschool education]. Galicia: Consellería de Cultura, Educación e Ordenación Universitaria.

10

GEOMETRY LEARNING OF CHILDREN IN DIGITAL ACTIVITIES

Ahmet Sami Konca, Sema Baydilli, and Elena Severina

Vignette

Asya is a preschool teacher at an elementary school. She teaches the only pre-school class at the school, which is located in the suburb of Kırşehir, Turkey, and whose students are generally from a low socioeconomic (SES) back-ground. The preschool classroom is located on the ground floor of the school along with the first-graders. There are three girls and five boys in her class, all aged four to five years. Even though not all the children attending the school are of low SES, all the children in Asya's class are.

Asya mostly follows the national curriculum, which emphasises play-based activities. She is also open to using technology in her classroom. The class-room is equipped with a smartboard, a desktop computer, and three tablet computers. Asya occasionally brings along mobile devices for the children to use. Only one child has access to digital technologies beyond a television at home. The children were found to interact willingly with digital technologies when Asya gave them the chance.

Asya had taught basic geometric shapes before. She aimed at teaching the children to recognise and learn the fundamental features of the shapes. First, the children learned songs about the features of a triangle, square, rectangle, and circle. The lyrics included names as well as the number of sides and corners of the shapes.

Asya and three of the children (Sude, Ela, and Cihan) sat around a table in the classroom. She handed a tablet computer to each of the children. There were pictures of many objects on their screens, such as a book, an envelope, a coin, some pita bread (a traditional Turkish bread), a door, a clock, a ball, a traffic sign, and a dartboard (Figure 10.1). There were also four boxes, one located in each corner of the screen, into which the children were to sort the

FIGURE 10.1 Screenshot of the activity

objects according to their shape. If the children attempted to slide an object into a box pertaining to another shape, the object would simply not be able to be dropped into the box. The children took their tablets and began aimlessly sliding the objects around.

"Ela, do you remember the songs we learned this week?" Asya asked. "Yes, I learned about triangles," Ela replied. Suddenly, Sude said, "Cihan, look, I found some pita bread." Cihan looked at the screen of her tablet and said, "It's like a pizza." Asya then asked, "Cihan, how is it like a pizza?" to which Ela replied, "It is like a pizza, yeees." Asya opened up a photograph of a pizza on the smartboard. Sude pointed excitedly, "Look, it has a round shape. And this pita bread is round too!" Cihan went over to the smartboard, drew a circle around the pizza, and said, "Look, Miss, it's like a pita bread. Both are round."

Asya was aware that the children tended to label the circle[1] as "round" rather than "circle." Therefore, Asya ensured that the children used the correct word, "circle."

"Cihan, do you remember the song about a circle?" Asya asked. "Yes, these are circles," Cihan replied. Asya then asked, "What were the other songs about?" The children all responded, saying, "Triangles, squares, and rectangles." Asya then asked the children, "Okay, now look at your tablets. Can you find those shapes?"

Asya aimed to improve the children's awareness in matching real-life objects with the correct shapes. She also used the definitions of the shapes during the activity.

Asya asked, "Look, there is a door, a coin, a book, and some other objects. Now we will play a game. Who can find the objects which have three corners and three sides?" The children began sliding the objects around. While Ela and Sude tried to count the corners of some objects, Cihan silently looked at the screen. Then he turned to the other children and pointed to a traffic sign. Cihan stated, "This traffic sign, yes. See, it has three corners." Suddenly Sude piped up with, "And this is a triangle piece of cheese!"

The children easily found the triangles. They showed a mountain, a hat, a roof, and a corn chip as triangular objects.

Then, Asya asked the children to find squares. "Okay, children," Asya said to get the children's attention, "Which of the objects have four corners and four equal sides?"

The children had difficulty in differentiating the squares from the rectangles. They talked about what they thought were squares. However, as they were a little confused, Asya repeated the definition of a square for them.

"How do we introduce the square?" Asya asked, "Think, I'm a square, how many sides do I have?" to which Ela replied, "Four sides, with two sides short and two sides long." Cihan thought differently, saying, "Noo, same sides."

The children notably used "same" to mean "equal." Asya then showed them a square button.

"Look, this is a square. How many sides does it have?" asked Asya. Ela answered first, "Four sides," then Sude said, "Four sides, same sides." Asya then asked, "Can you show me the sides, Sude?" Sude pointed to the sides using her finger, and said, "This is a square." Asya then continued, saying, "Okay, let's find some more squares."

The children discussed this as they looked for the squares. They created dyadic and triadic interactions and showed what they had found.

Ela pointed to a square piece of chocolate and said, "This has four corners," and Sude added, "Same sides too." Asya agreed, saying, "Yes, the sides are the same length," and then, "So, is it a square?" Sude replied, "Yes, but it is different."

In fact, the chocolate had only been rotated; hence, it appeared to be different to Sude. Once the children had found the squares, it was time to consider the rectangles. Asya thought she could start with a real object.

"What is the shape of this smartboard, children?" asked Asya. Cihan looked at the smartboard for a second and said, "It's a rectangle." Then Asya asked, "Why do you think that, Cihan?" Cihan pointed to the sides of the smartboard and said, "This one is long, and this one is short." Asya then asked, "So, how many corners does it have?" Both Ela and Sude replied excitedly, "Four!" Then, Ela pointed to a picture of a towel on her tablet and said, "Look, this is my dad's towel. It has two long sides and two short sides." "Well done, Ela," said Asya, "Now, let's sort the shapes into the boxes. First, slide the circles into the green box."

The children easily moved the circles into the correct box. Then Asya asked the children to collect the shapes with three corners and three sides. The children collected them into another box. However, when it came to the rectangle shapes, they sometimes confused the squares with the rectangles again. When they encountered a rotated object, they had difficulty in properly recognising it.

Cihan's finger was on a rotated rectangular portrait. Ela looked at his screen. She pulled his finger aside and said, "That's a square, so move it over there." Cihan followed her direction. However, the object did not fit into the box. "Let me try," said Ela. She tried but was also unsuccessful. Asya then asked, "Ela, Cihan, how many sides does it have? Check it." "Four," replied Cihan. Ela looked at the screen more closely and then said, "They are different. Two of them are long, and two are short." Asya then asked, "So, which shape is it?" to which Cihan proudly stated, "A rectangleee!"

Reflective questions

1 Young children often only experience standard geometrical shapes in stereo-typical ways, with no rotations or reversals. How can educators avoid this?
2 What do you think about the ways in which digital technology is used in the learning event depicted in the vignette?
3 How do second-hand experiences with iconic representations (i.e. pictures) of objects contribute to children's conceptual understanding of geometric shapes, and which role do first-hand experiences with real objects play in this process?

Commentary 1: geometry learning of children in digital activities

Author: Ahmet Sami Konca

In Turkey, early childhood education is centrally organised, with the main policymaking body being the Ministry of National Education (MoNE). MoNE implemented the latest early childhood education curriculum in 2013 (MoNE, 2013). The curriculum is designed for children aged 36–72 months old and aims to provide enriched learning environments in order to prepare them for elementary education. It also focuses on all aspects of their development, and therefore is defined as a developmental programme. The curriculum has two key components. The first is an eclectic model (Zais, 1976), in that teachers can apply different methods during the implementation of the curriculum. Second, the spiral approach (Bruner, 1960) involves repetition of previously learned knowledge through reinforcement with new knowledge.

The early childhood education curriculum bears the traces of Bishop's six fundamental activities (Bishop, 1988a). It mentions mathematical concepts – such as numbers, patterns, geometric shapes, sorting-comparing-grouping, measuring, and spatial sense – for the cognitive development of children. It is noted that teachers can also add concepts or objectives that are not mentioned in the curriculum if needed. While designing activities and daily plans, teachers take responsibility for aligning the objectives with the children's developmental capabilities. The curriculum strongly advocates that activities should be play-based and child-centred, as well as hands-on and engaging. The curriculum also underlines that teachers should consider the environmental conditions in which a child lives. It refers to the culture of children by remarking on the environmental conditions. Therefore, teachers must take into account the objectives and development of the children, as well as the culture in which the children will grow up.

The vignette was derived from within a project, which focused on teachers' integration of digital technologies into early mathematics education. The activity presented in the vignette was mainly related to designing and locating, among the six fundamental activities, which were discussed by Bishop (1988b). For the activity, the teacher considered the objectives as: (i) identifying names, (ii) recognising basic properties of geometric shapes, and (iii) matching real objects with geometric shapes. The teacher included squares, rectangles, triangles, and circles in the activity, and employed the use of tablet computers in the digital activity.

The activity can be evaluated via four aspects: objectives, developmental capabilities of children, consideration of the environmental conditions, and the use of digital technologies in the activity. While the first three aspects are mentioned, there is notably no statement about digital technologies within the curriculum. However, the eclectic approach of the curriculum may cover the application of digital technologies during the curriculum's implementation.

Objectives

The teacher aimed at three objectives to be completed within the activity. First, the teacher had taught the fundamental features of the shapes, and then aimed at the children matching the features to the objects. Although the activity seemed to focus on the children matching real objects with geometric shapes, the teacher sometimes returned to the first two objectives. For example, when the children were unsuccessful in linking objects with the correct shapes, the teacher would remind them about their prior learning, and finally would provoke them to match it again. The teacher used shape definitions while referring to the children's prior learning during the activity. Therefore, the teacher aimed to provide consistent and accurate information in order to prevent misconceptions regarding the shapes. Constant reliance on prototype shape images can lead to misconceptions being formed. Therefore, the teacher also used rotated images as a means of preventing misconceptions being formed by the children. The teacher provided multiple representations of figures within different orientations. For example, when using the rotated square of chocolate, the children became aware of the rotation and labelled the chocolate as a different square.

Developmental capability

The developmental capability of children was another important aspect of the activity presented in the vignette. The teacher had to consider the children's cognitive and other developmental features while preparing the activity. To achieve this aim, the teacher tried to realise the objective of the activity in harmony with the children's developmental abilities. For example, the teacher instructed the children to "sort" the objects rather than draw geometric shapes. Additionally, the teacher asked the children to slide the objects into the boxes by using the touch screen of the tablets. Even though the teacher's aim was not related to the children's drawing of the shapes, the children sometimes imitated the drawing action when using their fingers to draw shapes around the objects. The children were able to use the touch screens of the tablet computers to drag the objects into the appropriate boxes more easily than using a mouse to click and drag.

Culture

The third essential aspect of the activity was the teacher's consideration of the culture of the children by designing the activity to be culturally appropriate. Therefore, the activity included elements appropriate to the children's cultural background. All children in the classroom belonged to the same cultural background. The objects used in the activity were considered to be familiar to those seen in the children's daily life. For example, one child recognised her father's towel and focused on that. Using objects familiar to the children might improve their awareness, which then could lead the children to thinking about the shapes of objects which they encounter daily. This can be explained by Vygotsky's (1986) perspective about the interrelationship between spontaneous everyday concepts and nonspontaneous scientific concepts. He proposed a rationale between the concepts that children encountered in daily experience and the concepts taught in school (Howe, 1996). In the vignette, when the children learned the definition and features of the shapes, they connected this new knowledge to their prior experience. On the other hand, the teacher encouraged the children to apply the new knowledge to concrete examples (that is, the teacher asked the children about the shape of the smartboard). The activity included another component of the culture: language. Although the teacher intentionally used familiar daily objects for the activity, the role of the language emerged instinctively and freely. As underlined as a footnote in the vignette, the children chose their words as they would in daily life. Notably, the children used "same" to mean "equal" when comparing the length of an object's sides. Moreover, some Turkish words for geometric shapes included veiled pointers. The syllable "üç" in "üçgen" (triangle) means three and the syllable "dört" in "dörtgen" (quadrilateral) means four, referring to the number of sides and corners of the shape.

Digital activity

Lastly, the vignette consisted of a digital activity. The term "digital activity" is used in this chapter to describe activities that employ digital technologies. In the vignette, the teacher provided a tablet computer to each of the children for use during the activity. The children interacted with each other and their teacher, and also shared their ideas. Dyadic and triadic interactions during the activity included expressing their thoughts and negotiation. The interactions were fundamental to the activity. According to Arnott (2016, p. 276), the negotiation between the child and the context can result in social experiences as:

i   reciprocal behaviours and interactions exhibited by the children (i.e. children's seeking out interactions, helping their peers, negotiating),
ii   social participation of the children (i.e. parallel and associative play), and
iii   social status roles and technological positions (i.e. children's ownership of devices, children's observing peers' screens as spectators).

Interactions also included scaffolding (Wood, Bruner, & Ross, 1976) which means that the children interacted with each other while the teacher made suggestions, asked probing questions, and encouraged the children.

Summary

In brief, the vignette included a digital activity, which consisted of four core aspects: the objective of the activity; the developmental capability of the children; the cultural background of the children; and the use of digital technology. In order to support geometry learning in children through digital activities, these aspects should be considered and enriched by parents and early childhood educators alike. The vignette showed that children may experience difficulty in learning geometric shapes. Focusing on the shapes separately, using multiple drawings, providing rotated shapes, and encouraging children to link shapes with real objects can improve the quality of such activities.

Digital technologies can be useful in early mathematics activities. However, the effective use of digital technologies should be taken into consideration. Its effective use can be described in the following way: "Effective uses of technology and media are active, hands-on, engaging and empowering, give the child control; provide adaptive scaffolds to ease the accomplishment of tasks, and are used as one of many options to support children's learning" (NAEYC & Fred Rogers Center for Early Learning and Children's Media, 2012, p. 6). It should also be underlined that the successful usage of digital technologies implies combining both digital and non-digital activities, rather than separating them. Digital and non-digital activities each have their own specific characteristics and are not substitutes for each other. Therefore, purposeful and appropriate use of digital technologies in early childhood education requires both the interaction of children with the context and the right balance between digital and non-digital activities.

Commentary 2

Author: Sema Baydilli

In Turkish early childhood education, mathematics education aims to contribute to the cognitive development of children, to foster a positive attitude towards mathematics in children, to help children connect with previously learned conceptual knowledge, and to understand why and how mathematical concepts are used. In addition, mathematical activities should aim at improving children's mathematical inquiry skills. Through applied mathematical activities, children should be able to recognise patterns around them, develop assumptions and try them out, solve problems, reason, and communicate using mathematical concepts. Mathematics should also be experienced through examples that children may encounter in daily life.

In Turkish daily preschool education programmes, mathematics activities take place at certain times of the day. Usually, these activities are teacher-directed or children work by themselves from a textbook. Although it is emphasised in the national preschool curriculum that activities should be interconnected, most teachers plan independent activities. The main reason for planning in this way is the ability of both parents and schools to observe and document children's learning more clearly. Occasionally, teachers also plan independent mathematics activities in order to produce more concrete data so as to meet the varied expectations of schools. Although the Turkish national curriculum does not include paper-and-pencil mathematics exercises or mathematics textbook-based activities, teachers prefer ready-made programmes due to their ease of application, both in public (state) and private preschools. Whilst such kinds of books should be used, the main content should ideally be based on the needs of the children. Generally, teacher-guided activities take three or four hours per week for the teaching of scheduled mathematics activities. In Turkish preschools, classes are separated by age (for three-year-olds, four-year-olds and so on), with the mathematics learning curriculum planned according to the class age level and their capability. For example, not all four-year-olds have the same learning level, but teachers generally apply a standard teaching style aimed at specific sub-goals of the curriculum. Therefore, there are different activities for each age level, starting from the youngest age (36 months) through to the end of preschool age (72 months), and the curriculum follows the principle "from simple to complex."

Related to the vignette, a typical mathematics activity begins with an evaluation of the children's background. Once the teacher has understood the readiness of the children, they start the activity with a narration, and then move on to paper-based activities. For example, a matching activity can begin by playing with toys in which the children have to form matches, and then it continues with a worksheet where the children are tasked with matching pictures. The teaching focus is mainly on the second part, the application of worksheet-based activities, rather than natural inquiry as to the meaning of matching. This can lead to memorised learning.

In this vignette, the children had previously learned about geometric shapes and had reinforced their learning through various activities. Since the teacher aimed for the children to remember the basic features of the shapes, the children were given a sense of shape through song and rhyme in which descriptions of the sides, corners, and shape were emphasised. In the subsequent part of the activity, the children were tasked with matching the shapes with pictures of daily objects.

When mathematics education relates to technology, it touches other domains of development and learning, and thereby attracts the attention of today's technologically enabled children. However, the point that should be considered here is whether technology should be used just to facilitate the discipline being taught, or to facilitate the usage of technology itself. In order to deal with this dilemma, teachers and educational programme developers sometimes try to integrate technology into events. In today's technological age, and with many applications and programmes in play, teachers can both enrich children's learning spaces as well as narrow them.

In the vignette, the teacher tasked the children with using tablet computers in order to match shapes to pictures of real objects. With a guiding question from the teacher, one of the children matched a picture in his mind of an actual object with a triangle, and matched an abstract concept previously learned with an image of a concrete figure. The teacher could have just shown the tablet with a simple instruction, saying, "Match the objects you see on the screen to similar shapes." However, the teacher allowed the children to recall a song they had previously learned, thereby enabling them to also recall prior knowledge. Making the right connections and associations with mathematical concepts ensures sustainability of the learned content.

When teaching children shapes, the first task is to develop in the child's mind a holistic picture of what the shape looks like, to learn what name it has, and to experience where and when it is used in daily life.

Even if the event in the vignette appears to be a technology-based activity, the activity does not wholly focus on technology per se. Technology was used as a tool in the activity, but teaching how to use a technological tool was not the primary goal. Teacher support should be sought for children to utilise the tablet as an objective in the activity. With this support, children can search among images that appear in the search engine, and among them, the children can discover images of related figures. In this way, they can develop more of an idea about the features and intended use of the technological tool they are using. In short, in the vignette, the technology was applied as a visual tool, but there was no aim for its utilisation to search for and reach additional information. The most basic feature of mathematics is to reach information by searching and questioning as, by its very nature, mathematics involves aspects of inquiry, exploration, and analysis (Clements, Sarama, & DiBaise, 2003).

In terms of the six basic mathematical activities put forward by Bishop (1988b), this vignette includes locating, designing, and playing, although not exclusively. The focus of the vignette is on the shapes, that is, designing. The children may have received information previously about the sides and corners of the shapes, but in the current activity, different concepts were also discussed, for example, the concepts of "more" and "less" when counting how many objects accumulated in the boxes in which they had classified the images according to their shapes. Mathematical activities should be afforded greater focus and should be taught in relation to the environment – especially for three- to six-year-old children. If this is not applied, activities will remain in the memory only temporarily and cannot then be transferred to long-term memory.

Games and activities in preschool education programmes help children understand spatial relationships such as location, movement, and distance. Children begin to define spatial relationships in drawings, models, and maps. In the preschool period, working with simple maps helps children manipulate space-related words and objects according to various positions, and thereby develop spatial logic. For example, children can try to locate a toy in their classroom with the aid of a map provided by their teacher. The map should include directional words, signs indicating geometric shapes, distance between signs in steps, and a pictorial representation of the toy they are tasked to find. As an example, in the vignette, the teacher could have organised a shape map that included some direction. Instead, the teacher used a tablet including different pictures of objects, which were related to shapes. In this way, the teacher enabled the children to reach the concept of shape by thinking differently, not directly but indirectly.

At the end of the event reported in the vignette, the children would be able to draw shapes or create a tangram consisting of different shaped pieces. Thus, the activity could move to a new dimension.

Finally, the children were unsure when it came to distinguishing squares and rectangles. It is very normal for these two shapes to be mixed in children of this age because they can count four edges and call them squares, but after gaining experience for a certain amount of time, they can grasp the rectangle concept by comparing the lengths of the edges, which is an activity level in van Hiele (1986). The children needed orientation in order to distinguish whether the given object was square or rectangular. On a topic that children do not know or have doubts about, teachers should arrange the learning environment accordingly for the children to reach the right information by means of different methods. Whilst it is useful for children to discuss and to try to understand a situation through examples, instead of the teacher asking questions through simple verbal instruction, the teacher can describe the distinction between squares and rectangles through more active means. In the vignette, the teacher tried to understand the confusion with the question she originally asked the children. However, she was able to detect the confusion between the square and the rectangle with the question prompting the children.

Commentary 3

Author: Elena Severina

In Norway, early childhood education is from zero to six years old and, as in Turkey, is not compulsory. Four- and five-year-olds go to a preschool group in kindergarten. Norwegian early childhood education, governed by the *Framework Plan for Kindergartens – Content and Tasks* (Ministry of Education and Research, 2017), follows a social-pedagogical tradition. Play, both indoor and outdoor, is the main arena for learning and social and linguistic development. According to the Framework Plan, children's curiosity, creativity, and will to learn are seen as essential in forming learning processes, and therefore, should be acknowledged and stimulated (Ministry of Education and Research, 2017).

The Framework Plan describes mathematical content in its own chapter, "Quantities, Spaces and Shapes," and includes "play and investigation involving comparison, sorting, placement, orientation, visualisation, shapes, patterns, numbers, counting and measuring" as well as "asking questions, reasoning, argumentation and seeking solutions" (Ministry of Education and Research, 2017, p. 53). The chapter also provides a list of the explicit expectations of kindergartens regarding what mathematical learning opportunities the children should be provided with, and how this should be done. Although this is not formally acknowledged, overall description and objectives of the learning area "Quantities, Spaces and Shapes" are in coherence with the six fundamental mathematical activities (Bishop, 1988a).

In the digital activity described in the vignette, children use basic properties of the geometrical forms to sort a set of pictures of the real objects by their shape into four boxes labelled with drawings of circle, triangle, square and rectangle. In some cases, children are also challenged to explain their choices. In Norway, these would fit the objectives,

> By engaging with quantities, spaces and shapes, kindergartens shall enable children
> - to investigate and recognise the characteristics of different shapes and sort them in a variety of ways […]
> - to use their bodies and senses to develop spatial awareness
> […]
> Staff shall use books, games, music, digital tools, natural materials, toys and equipment to inspire children's mathematical thinking.
> *(Ministry of Education and Research, 2017, pp. 53–54)*

According to the literature used in early childhood education in Norway, to develop children's mathematical thinking within geometry, kindergartens should create learning environments where children primarily focus on analyses of the shapes' properties, how shapes can be composed and decomposed, and develop understanding of patterns (Carlsen, Wathne, & Blomgren, 2017; Nakken & Thiel, 2019; Solem & Reikerås, 2017). Also, as mathematical conversation stimulates children's curiosity and language development by providing them with an oppor-tunity to express their own mathematical ideas and elaborate them (Carlsen, 2016; Skorpen, 2012), it has to be a part of such activities. In Norway, recycled and natural materials, Lego, Magformers, pattern blocks, tangrams, and so on are interchangeably used with cameras, smartboards, apps, and other information and communications technology to stimulate geometric thinking. Following the Norwegian Framework Plan, kindergarten staff are expected to use "digital practice" (Ministry of Education and Research, 2017) as a working method, while the curriculum in Turkey does not mention the digital technologies explicitly.

Designing, one of the fundamental mathematical activities, is about the concep-tualisation of objects and artefacts, and here the idea of "shape" is central (Bishop, 1988a). Learning about shapes is unavoidably related to the idea of classification of the shapes by properties as well as argumentation for such, which belongs to explaining fundamental activity (Bishop, 1988a). To classify an object by its shape, one also needs to locate, orient, and perform some mental transformations in real or digital environments. Such skills belong to the fundamental activity *locating* (Bishop, 1988a). Therefore, when learning about shapes, designing, locating, and explaining are naturally intertwined. This is also evident in the vignette as children name the shapes and their properties ("and here is a *triangle* piece of cheese!"), support their hypotheses about shapes classification with explanations ("Cihan pointed to the sides of the smartboard and said, 'this one is long, and this one is short'"), and locate and move a certain object to the corresponding box ("it's a square, so *move it over there*").

Van Hiele (1986) describes the development of geometric thought as a con-tinuous process consisting of five sequential age-independent levels, starting from merely recognising the geometric shape ("visualisation") to being able to write a formal proof ("rigour"). Clements and Battista (1992) proposed the existence of a "pre-recognition" level, when only a subset of the visual characteristics of a shape is noticed, resulting in an inability to distinguish between geometric shapes. The developmental process depends on a variety of experiences one gains as well as instruction language (van Hiele, 1986; Clements & Battista, 1992).

In order to proceed from prototype-based thinking typical for the visualisa-tion level to properties-based thinking, one needs to discriminate properties of the shapes and realise that they are essential for classification. Following Hana (2012), the defining properties (invariants) of the geometric shape stay the same as long as the shape belongs to the same class, while other properties (variations) will change within their dimensions of the variation. Creating a learning environment in kin-dergarten involves opening up to possible dimensions of variation; looking at the kind of variation and invariance present in the situation can provide a clue to poten-tial learning (Hana, 2012). Van de Walle, Karp, and Bay-Williams (2015) state that for development of properties-based learning, it is important to get various prac-tical experiences with rotation, scaling, assembling, or decomposing of the shape as well as transformations from two-dimensional to three-dimensional. Findings of Berthelot and Salin (1998) suggest that variation of the space size (macrospace, mesospace, and microspace) where children get geometric experiences is important to develop spatial awareness and geometric thought. In the kindergarten context, a microspace could be a board game or a piece of paper, a mesospace could be a room or playground at kindergarten, and a macrospace is the unknown forest or city (Føsker, 2012).

The conversation in the vignette builds around children's experiences with a sorting game on a tablet. Although the core of the learning environment is in the microspace of the game, where both iconic and symbolic representations (Bruner, 1966) of the real objects are used, the teacher successfully extends it to mesospace of the classroom when inviting the children to explore the properties of a pizza image and later a smartboard. The shift from micro- to mesospace allows the children to use more concrete representations of circles and rectangles and to explain why they mean the object has a certain shape.

When the children are having a mathematical conversation (Fosse, 2016) about the pita bread, they compare it with pizza: objects are mathematically similar although neither is perfectly "round." By posing the open question "How is it like a pizza?" and having pictures of both objects available simultaneously, Asya helps the children separate variances of the objects from the common invariant (roundness), which is evident in Cihan's and Sude's bodily and verbal explanations.

Further, Asya uses a real object, the smartboard, to talk about rectangles. Here, by asking, "Why do you think that?" Asya invites Cihan to justify the statement that the shape is a rectangle. Cihan's mathematical explanation (Yackel, 2001) regarding side length and Asya's follow-up question about corners helps the children identify one rectangle in the digital environment of the game.

Based on the dialogues about smartboard and portrait, both Ela and Cihan seem to have skills typical to the transition from the first (visualisation) to the second (analysis) of van Hiele's levels. Children started to identify classes by the properties, but they still need some more experience to develop the skill of mental transformation (rotation) of the objects.

Figure 10.1 shows a screenshot of the game. All objects can be seen as four groups: triangles, circles, squares, and rectangles. Objects in each group vary by size, orientation, and colour, but the qualities of these variations are different. Six objects with a triangular shape are equilateral or isosceles, and only one triangle is rotated from the norm. Two objects have a circular shape, and one is almost a circle. Both[2] objects with a rectangular shape and one of two squared objects are rotated. What can be said about potential learning? As children at the pre-recognition level would not differentiate squares from rectangles (Clements & Battista, 1992), and at the visualisation level (van Hiele, 1986), would still use prototype-based reasoning, classification of the rotated objects will be challenging. Taking into consideration variation within each group of objects, the circle and the triangle should be the easiest to recognise, while the rectangle would be the most challenging. Circles and rectangles are few in the microspace. This could be the reason why the use of the examples from mesospace was necessary. The learning environment could be used to pose questions like "What makes a circle a circle?" "What makes this nacho look different from other triangular objects?" "What do all these four-sided shapes have in common? What makes them different?" and so on.

In the vignette, children are invited into the digital learning environment as they engage with a geometry sorting game on the tablet. This environment has affordances different to a dynamic geometrical environment (DGE) as the objects can only be moved to the destination point, not rotated, changed, or scaled. Sinclair (2018) discusses features of DGE that could be helpful in the transition from prototype-based to property-based reasoning already in kindergarten. DGE allows the experiencing of a continuous variation of the shape while preserving the class of the shape. This produces many examples of non-prototypical geometric shapes and highlights its invariants. For example, dragging a vertex of a triangle will continuously modify its shape, while keeping it a three-sided polygon. Therefore, exploration of the shapes in DGE supported by ongoing discussion could be beneficial for young children in kindergarten to start to describe and define basic geometric shapes (Sinclair & Moss, 2012).

Development of property-based reasoning in preschool children is dependent on varying experiences with geometric shapes. In talking about digital methods, I would suggest that explorative activities in DGE, such as GeoGebra or Sketchpad, together with supportive mathematical conversations, might help children identify variances and invariants of shapes. While exploring, it could be a good idea to do one change at a time (rotate, scale, move one of the vertices, and so on) and always check whether the shape still belongs to the same class.

Summing up, the development of geometric thinking depends on the variation of the experiences and instructional language, and not just age. Children should be active when experiencing variation in both sizes and types of the learning environments, representations used, transformations, and contexts available.

Notes

1 In Turkish, while *daire* means circle, *yuvarlak* is used to imply a round shape. Moreover, *yuvarlak* tends to be used instead of *daire* for circle in daily language. However, as the meanings of *yuvarlak* and *daire* are not actually the same, Asya ensured that the children correctly used *daire* when naming a circle.

2 Here, in alliance with the vignette and the other commentaries, I do not count squared shapes as rectangular to be in coherence with the children's level in van Hiele's model.

References

Arnott, L. (2016). An ecological exploration of young children's digital play: Framing children's social experiences with technologies in early childhood. *Early Years, 36*(3), 271–288.

Berthelot, R., & Salin, M. H. (1998). The role of pupils' spatial knowledge in the elementary teaching of geometry. In C. Mammana and V. Villani (Eds.), *Perspectives on the teaching of geometry for the 21st century,* (pp. 71–78). Dordrecht: Kluwer.

Bishop, A. J. (1988a). *Mathematical enculturation: A cultural perspective on mathematics education.* Dordrecht: Kluwer.

Bishop, A. J. (1988b). Mathematics education in its cultural contexts. *Educational Studies in Mathematics, 19*(2), 179–191. Retrieved from https://link.springer.com/article/10.1007/BF00751231

Bruner, J. S. (1960). *The process of education.* Cambridge, MA: Harvard University Press.

Bruner, J. S. (1966). *Towards a theory of instruction.* Cambridge, MA: Harvard University Press.

Carlsen, M. (2016). Matematiske samtaler i barnehagen: Utfordringer og muligheter. [Mathematical conversions in kindergarten: Challenges and opportunities.] In R. Herheim, and M. Johnsen-Høines (Eds.), *Matematikksamtaler. Undervisning og læring – analytiske perspektiv* [Mathematical conversations. Teaching and learning – analytical perspective] (pp. 221–239). Bergen: Casper Forlag.

Carlsen, M., Wahtne, U., & Blomgren, G. (2017). *Matematikk for barnehagelærere* [Mathematics for kindergarten teachers] (3rd ed.). Oslo: Cappelen Damm AS.

Clements, D. H., & Battista, M. T. (1992). Geometry and spatial reasoning. In D. A. Grouws (Ed.), *Handbook of research on mathematics teaching and learning,* (pp. 420–464). New York: Macmillan.

Clements, D. H., Sarama, J., & DiBaise, A. M. (2003). *Engaging young children in mathematics: Standards for early childhood mathematics education.* New York: Routledge.

Fosse, T. (2016). What characterises mathematical conversations in a Norwegian kindergarten? *Nordisk matematikkdidaktikk, 21*(4), 135–153.

Føsker, L. I. R. (2012). Grip rommet! Barns utvikling av romforståelse og barnehagelærerens systematiske arbeid med det [Catch the space! Children's development of spatial understanding and kindergarten teachers' systematic work with it]. In T. Fosse (Ed.), *Rom for matematikk – I barnehagen* [Space for mathematics – in the kindergarten], (pp. 61–89). Bergen: Caspar Forlag.

Hana, G. (2012) Varians og invarians [Variation and invariance]. In T. Fosse (Ed.), *Rom for matematikk – i barnehagen* [Space for mathematics – in the kindergarten], (pp. 23–42). Bergen: Casper Forlag.

Howe, A. C. (1996). Development of science concepts within a Vygotskian framework. *Science Education, 80*(1), 35–51.

Ministry of Education and Research, (2017). *Framework plan for kindergartens – content and tasks.* Norwegian Directorate for Education and Training. Retrieved from www.udir.no/globalassets/filer/barnehage/rammeplan/framework-plan-for-kindergartens2-2017.pdf

Ministry of National Education (MoNE). (2013). *Preschool curriculum.* Turkey: Talim Terbiye Kurulu Başkanlığı.

NAEYC & Fred Rogers Center for Early Learning and Children's Media. (2012). *Technology and interactive media as tools in early childhood programs serving children from birth through age 8.* Joint position statement. Washington, DC: NAEYC; Latrobe, PA: Fred Rogers Center at St. Vincent College. Retrieved from www.naeyc.org/content/technology-and-young-children.

Nakken, A. H., & Thiel, O. (2019). *Matematikkens kjerne* [The core of mathematics], (2nd ed.). Bergen: Fagbokforlaget.

Sinclair, N. (2018). Time, immersion and articulation: Digital technology for early childhood mathematics. In I. Elia, J. Mulligan, A. Anderson, A. Baccaglini-Frank, and C. Benz (Eds.), *Contemporary research and perspectives on early childhood education,* (pp. 205–211). Heidelberg: Springer.

Sinclair, N., & Moss, J. (2012). The more it changes, the more it becomes the same: The development of the routine of shape identification in dynamic geometry environments. *International Journal of Education Research, 51 & 52*, 28–44.

Skorpen, L. B. (2012). Utforskende tenkning og samtale. Filosofiske samtalar om matematiske spørsmål. [Exploratory thinking and conversation. Philosophical conversations about mathematical questions] In T. Fosse (Ed.), *Rom for matematikk – i barnehagen* [Space for mathematics – in the kindergarten], (pp. 43–60). Bergen: Casper Forlag.

Solem, I. H., & Reikerås, E. K. L. (2017). *Det matematiske barnet* [The mathematical child] (3rd ed.). Bergen: Caspar Forlag.

van De Walle, J. A., Karp, K., & Bay-Williams, J. M. (2015). *Elementary and middle school mathematics: Teaching developmentally, Global Edition*, (9th ed.). New York: Pearson.

van Hiele, P. M. (1986). *Structure and insight. A theory of mathematics education.* Orlando: Academic Press.

Vygotsky, L. (1986). *Thought and language* (A. Kozulin, Trans.). Cambridge, MA: MIT Press.

Wood, D. J., Bruner, J. S., & Ross, G. (1976). The role of tutoring in problem solving. *Journal of Child Psychiatry and Psychology, 17*(2), 89–100.

Yackel, E. (2001). Explanation, justification and argumentation in mathematics classrooms. In M. v. d. Heuvel-Panhuizen (Ed.), *Proceedings of the 25th PME International Conference* (Vol. 1, pp. 9–24). Utrecht, The Netherlands: Utrecht Freudenthal Institute, Utrecht University.

Zais, R. (1976). *Curriculum: Principles and foundations.* New York: Thomas Y. Crowell.

PART 6
Playing

11

"THIS IS THE SAFE. IT HAS A NUMBER AND NO ONE ELSE KNOWS IT"

Playing with mathematics

Maulfry Worthington, Elizabeth Carruthers, and Lone Hattingh

Vignette

The research setting in which this vignette was documented, is a maintained nursery school in an inner-city district of a large multicultural city in the southwest of England.[1] The nursery school welcomes families from many different ethnic backgrounds, and at the time of data collection, children attending the nursery spoke 13 different languages. The children in this observation are three to four years of age and in their final year of nursery school. Both boys' first language is English.

The vignette focuses on an episode of social pretend play in which the boys' collaborative dialogue grew from Isaac's personal interest in security safes, one of his many *funds of cultural knowledge* (Moll et al., 1992; Worthington, 2018; Worthington & van Oers 2016). During the course of their play, the boys made reference to many aspects of number and quantity, culminating in communicating ideas through emergent mathematical inscriptions (Carruthers & Worthington, 2006).

Throughout the year, Isaac initiated and engaged in pretend play more than any other child in the study, his play often highly complex and sustained. Isaac's interest in security technologies such as padlocks, security cameras, and safes arose from his first-hand experience at home and through his father's building work. He also has well-developed knowledge of money though his father's new work with his microbrewery, involving deliveries, invoices, payments, and counting cash, all activities in which Isaac has been involved. Shortly before this observation – and knowing of Isaac's interest in safes – his teacher Emma had brought in a small safe for the children to explore.[2]

The vignette is drawn from qualitative data gathered for case studies for doctoral research, investigating children's beginnings and development of their mathematical meaning making and their emergent understandings. This research underscores the potential of pretend play for *cultural-conceptual* mathematical learning and is founded on a Vygotskian cultural-historical and social-semiotic perspective (Vygotsky, 1978; Worthington & van Oers, 2016). Longitudinal, ethnographic data were gathered during the course of one year. These data comprise written observations of children's behaviours, talk, and learning, and children's inscriptions, all the children's explorations relating to their personal cultural knowledge. The following narrative is substantively taken from Worthington and van Oers (2016, pp. 58–59).

Jayden and Isaac decided to transport wooden blocks on a trolley. They moved a small cupboard that stood nearby to create a "safe," and then placed a keyboard and clipboard on top of it.

When another child came and took away one of the blocks, Jayden noted this down on his clipboard by writing wavy lines. He took the piece of paper off the clipboard and placed it in their safe, then tapped several keys on the keyboard, repeating this process every time a child removed one of the blocks.

Isaac announced, "This is the safe. There's a key, only one – and you press it here and it opens. It has a number and no one else knows it – one, one, eight, seven, zero, six. It's rather difficult to remember."

Jayden brought over some coins and cheques to put in their "safe." Isaac found a calculator and stuck it onto their safe with tape, explaining, "You need to press the buttons to get in the safe … it's four, nine, seven, nine."

Jayden pressed some numbers, making a "beep, beep, beep" noise as he opened the safe. Closing the doors Jayden asked, "What's the closing number?" Then said, "one, nine, five, two" as he pressed buttons on the calculator.

Later Jayden told his teacher Emma, "You need to give me one, nine, five, two." Emma explained that she did not have that amount in cash but could write a cheque and Isaac replied, "I need hundreds of pounds in cash!"

Emma managed to find a selection of coins in her purse and gave them to Jayden, who said, "Okay! We need to fill the box for one, five, six, zero pounds."

"It's a skip!" Isaac said excitedly as the box filled, "It's getting too full! It's more than one, five, six, zero pounds!" Isaac asked his teacher to write "one, five, six, zero pounds" on the box, and looking at the numbers and symbols, he took great interest in what she had written.

After several days creating and playing with their safe, Isaac used it to store "important information." He decided to write down the number of blocks being removed from the block area, counting, "One, two, three, gone! Gotta write it down and put it in the safe."[3]

Reflective questions

1   How can teachers ensure that they understand and value children's ". cultural knowledge" in relation to mathematics?
2   How might we support open opportunities for rich mathematical learning though play, particularly in settings where there are "downwards pressures" from school, for more formal approaches to mathematics?
3   Provide an example of another pretend play situation you have observed, and discuss how professional practice can support meaningful mathematics learning embedded in that play.

Commentary 1: pretend play as a context for mathematics

Author: Maulfry D. Worthington

The head teacher of this nursery school is committed to ensuring a democratic and open ethos, staff valuing and supporting children's self-initiated ideas, choices, and decisions, and how they express them. Mathematics and graphicacy (that is, children's own drawings, maps, writing, and mathematical inscriptions) have a very high profile and are valued and sensitively supported by staff.[4] Especially noticeable is the staff members' views of the child as "a *rich child,* active, competent and eager to engage with the world" (Dahlberg, Moss, & Pence, 2007, p. 7; italics in original), sharing a belief that how we construct our views of the young child and of early childhood, matters. Rather than narrow curriculum goals, with knowledge and skills transmitted *from* adults *to* children, learning is understood as social, cultural, and collaborative. Teachers encourage children to initiate ideas, valuing and supporting their complex thinking through rich dialogue and emphasising the intellectual over the academic, and children's meaning making over school readiness: this philosophy is the headteacher's.

Free and spontaneous play is an essential feature of this nursery school. Pretend play is neither themed nor planned by adults and is a popular choice for many of the children, their impromptu play frequent and often complex. Hewes (2014, p. 280) writes that for children, "adaptability, control, flexibility, resilience and balance result from the experience of uncertainty, unpredictability, novelty and non-productivity. These [are] essential dimensions of young children's spontaneous free play." Such play also allows children to be responsive to diverse aspects of their cultural knowledge within their play narratives, including their mathematical knowledge (Worthington & van Oers, 2016). Others who have investigated young children's social pretend play and the mathematics arising within it include Carruthers and Worthington (2006; 2011), Cook (2006), Munn and Kleinberg (2003), Munn and Schaffer (1993), van Oers (2010, 2013), and Worthington (2018; 2020).

In this nursery school, children self-initiate their mathematical ideas through their pretend play indoors and out, also instigating their own ideas and choices in open, adult-led small groups. In contrast to some of the previously published research into pretend play and mathematics by other researchers (e.g., Gifford, 2005), the children often choose to explore and communicate mathematical ideas freely through talk and graphical inscriptions within their rich play narratives. The level of the children's achievement in mathematics at this nursery school (relating to national assessment procedures and government inspections) has been repeatedly found to exceed expected outcomes for children of this age.

Danniels and Pyle (2018) point out that "play-based learning is distinct from the broader concept of play" (p. 1). They suggest first, that the term *play-based* encompasses children's own free play, initiated and developed by the children themselves, "voluntary, internally motivated, and pleasurable." However, rather than the common use of the term "fun" when referring to children's play, our observations of young children's spontaneous and free pretend play is that it can be both enjoyable, and, at the same time, also deeply serious and engaging for them.

A second interpretation of play-based learning identified by Danniels and Pyle (2018, p. 1) is of *guided play*, its most distinctive feature "who has control over the play activity." In England the term play-based is used only in teacher-planned (and often teacher-led) activities with specific learning goals in mind, and often provides very little opportunity for children to freely make choices in that play. In this vignette, Isaac and Jayden had complete ownership of their play in respect of the location in which they chose to play, the friend with whom they played, the resources and materials they chose to use, the focus and development of their play, and its duration. Our argument is that it is only children's truly spontaneous free play (within an open and supportive learning environment) that embodies important features of play. Such play allows children to explore their existing cultural knowledge, and can lead to rich learning.

The *Statutory Framework for the Early Years Foundation Stage* (EYFS), (DfE, 2017) covers the period from birth to five years, but in reality, almost all four-year-old children are in a "reception class" in a primary school.[5] One of the mandatory areas of the curriculum is mathematics, delineating "counting, understanding and using numbers, calculating simple addition and subtraction problems; and to describe shapes, spaces, and measure" (p. 8). The curriculum advises "play is essential for children's development, building their confidence as they learn to explore, to think about problems, and relate to others. Children learn by leading their own play, and by taking part in play which is guided by adults" (p. 9). However, although play is officially acknowledged, it may be fleeting and seldom includes children's mathematical explorations (Gifford, 2005).

The narrowing of the curriculum requirements and the "school readiness" agenda have resulted in a reduction of rich and sustained pretend play in most settings, and play is increasingly marginalised.[6] In common with many countries, England appears to share in an escalation of prescribed "skills-based" teaching, often resulting in restricted understandings of children's own mathematics and only limited opportunities for rich play. The increasing politicisation of education risks losing sight of young children's needs and development (Worthington, 2020). Citing Moss (2013, p. 9), Robert-Holmes (2015, p. 11) concludes "complex holistic child-centred principles, sensitive pedagogies and assessments [are] in danger of being marginalised [and have] the potential to reduce the rich competent child (and teacher) to a 'measurable teaching subject'" (Ball and Olmedo, 2013, p. 92). One of the most worrying outcomes of the pressures of "schoolification" of early years education, is that taught mathematical "skills" are often unrelated to children's development and existing cultural knowledge.

Culture and mathematics

Locating two ways of appropriating cultural knowledge, Vygotsky (1978) highlighted how children do so directly through their experiences in cultural and social situations and practices (leading to spontaneous, "everyday" concepts), and through instruction (leading to schooled, "scientific" concepts). For Vygotsky, pretend play helps in "bridging" these concepts (Worthington & van Oers, 2016, p. 53). It offers potentially rich social contexts that situate learning, allowing children to explore and draw on their existing cultural knowledge of mathematics (Worthington, 2018). Bishop (1988b) argued that, in the past, "the conventional wisdom was that mathematics was 'culture-free' knowledge" (p. 146). Identifying playing as one of his "universal activities of mathematics," Bishop (1988b) proposed that

> This may seem initially to be a rather curious activity to include in a collection of cultural activities relevant to the development of mathematical ideas… *It is even more important to include it, when considering mathematics from a cultural perspective. One is forced then to realise just how significant 'play' has been in the development of culture.*
>
> (pp. 149–150; italics added)

Mathematics in pretend play

In England, the term "play" is commonly used to refer to a variety of child-initiated activities such as puzzles, play with sand and water and blocks and bricks, and, whilst these have value for mathematics, for Vygotsky (1978, p. 96) pretend play is the "leading activity" for young children, its influence on their development "enormous." Play scholars acknowledge play as led by and belonging to children (van Oers, 2013). However, Gifford (2005) found many researchers in England identified a striking lack of mathematics in pretend play. Paradoxically, Worthington and van Oers (2016) found that play that is *owned by* children (rather than planned by adults, and without adult goals for mathematics), is frequently rich in contextually meaningful mathematics through their talk, behaviours, and graphical representations, their play narratives arising freely and spontaneously and shaped and developed by them. In this instance, the mathematical concepts the boys explored included *number* and *quantity, zero, money, and capacity*, all explored in a relevant context that grew from and was related to their own experiences and interests.

In their analysis of three- to four-year-old children's pretend play, Worthington and van Oers (2016) identified connections between the home cultural knowledge they brought to their play narratives and the mathematics they explored in their play. Their findings revealed also the extent of the children's spontaneous mathematical explorations within their pretence, as the vignette shows. Underpinned by Vygotsky's (1978) cultural-historical (or sociocultural) theory, the research on which this chapter is based regards concepts developing in tandem through interactions with cultural systems.

Playing in mathematics

Identifying six "universal" activities of mathematics, Bishop (1988a) specifies *counting, locating, measuring, designing, playing,* and *explaining*. In Bishop's terms, the mathematics the boys explored in this vignette included *counting, locating, designing, playing,* and *explaining*. Bishop (1991, p. 101) defines *playing* as "games; fun; puzzles; paradoxes; modelling; *imagined reality*; rule-bound activity; hypothetical reasoning; procedures; plans, strategies; cooperative games; competitive games; solitaire games; chance [and] prediction" (italics added), writing that, "once the play-form itself becomes the focus [...] then the *rules*, procedures, tasks and criteria [...] are the products of playing" (Bishop 1991, p. 45, italics added). However, rather than a feature of mathematical activity, play scholars regard "playing" as *a behaviour, an activity, and a context*. Children's spontaneous and free pretend play has the potential to embrace children's mathematical thinking. Seo and Ginsburg (2004) emphasise that adults need to "take the child's perspective, understand the child's current intellectual activities, and build on the mathematics to foster the child's learning" (p. 25). However, unlike Bishop's perspective, whilst children may be engaged in pretend play, there is no guarantee of meaningful mathematical learning (for example, see Gifford, 2005), but the play depends on the culture and philosophy of the setting, and the teacher's intentionality (Hewes, 2014). For additional means of consolidating and enriching children's mathematical understandings, see commentary 2.

Vygotsky (1978) showed how children's pretend play "depends on rules [...] it is the sense that children make of a particular cultural situation and its rules that motivates their actions" (p. 190). Vygotsky's sociocultural theory is exemplified well in this vignette. The play and learning are fully social, allowing exchanges of ideas through dialogue, actions, and the sharing of inscriptions: it is cultural in that its freedom permits the children to readily explore and extend their funds of mathematical knowledge. Bishop's (1988a) "imagined reality" also encompasses *rule-bound activity*, in which, according to van Oers (2013), the rules relate to "how [children] interact, how they use their tools, or how to organise the play" (p. 191).[7] Following Vygotsky, van Oers (2013, p. 191) identified three important features of pretend play, its *rules, degrees of freedom*, and children's *involvement*, features we use here to interpret aspects of the vignette. The rules of pretend play are shown in Table 11.1.

TABLE 11.1 The rules of pretend play taken from van Oers (2013, pp. 191–192)

Social rules (how to interact with one another)	The boys' turn-taking soon establishes them as equal partners in contributing ideas and cultural knowledge.
Technical rules (how to use a tool or piece of technology)	Isaac introduces the idea that the safe is opened with a key, at the same time referring to a "*difficult to remember*" number to open it.
	Isaac understands that some safes have a digital keypad to open and secure them. He fixes a calculator to the door of their "safe," explaining that its numerical buttons are to be pressed "*to get into the safe.*"
	Jayden makes a "*beep, beep, beep*" noise, perhaps associating their safe with other technologies that make this noise.
Conceptual rules (i.e. rules with a conceptual basis)	When another child removed a wooden block, Jayden showed he understood something of subtraction by "writing" down what had occurred, his wavy lines resembling adults' writing.
	The boys explore ideas of number strings as codes for opening and closing the lock.
	Isaac refers to "*zero*" three times, and Jayden once.
	Isaac understands that a long number with six digits will be "*rather difficult to remember.*"
	Jayden knows that valuables such as money are kept in safes.
	Jayden understands that a large amount of money would be needed to fill their box to capacity.
	Isaac understands that "*important information*" is kept in a safe."
Strategic rules (rules that support the course of an activity)	The boys know that they are free to use and move resources and furniture as they wish.
	Isaac is interested in how his teacher represents "*one, five, six, zero pounds.*"
	The strategic rules are implicit in the course of this play episode.

The boys' conceptual rules of mathematics include aspects of number (*subtraction, number strings, zero, money, quantity, graphical representations of quantity,* and *capacity*). Reflecting on her observation, their teacher wrote "both boys were using their experiences and understanding of numbers in a real situation [...] Jayden, in awe of large numbers in relation to money, knew that £156.00 was a large amount of cash" (Worthington & van Oers, 2016, p. 59).

Degrees of freedom

Van Oers (2013) writes that "the degrees of freedom" within play allow children to make their own choices concerning their "actions, tools and rules, etc." (p. 191). In contrast, many children's experiences are circumscribed by restrictions on play that reveals only, "*adults' perceptions* of children's interests, rather than children's authentic and immediate interests that have personal cultural meaning" (Worthington & van Oers, 2016, p. 52, italics in the original). Weisberg et al. (2015) refer to adult planned activities as "*disguised as play* [...] [such as] chocolate-covered broccoli" (p. 10) (italics added).

Levels of involvement

Deeply involved in their play, Isaac and Jayden continued to explore the same narrative on several subsequent days. Van Oers (2013) regards pretend play as an activity in which children are highly involved, and this level of involvement is evident in this vignette. In this nursery school, teachers value children's pretend play and support their meaning making through reflecting on their written observations; they engage in collaborative dialogue with them, and frequently model mathematical signs and representations in meaningful contexts. This professional community of practice appears to create a learning "hot spot," children instigating pretend play and exploring meaningful mathematics embedded in that play[8].

Commentary 2: the mathematics teacher as key player

Author: Elizabeth Carruthers

I am the head teacher of the nursery school and National Teaching School in this vignette; therefore, my commentary is from a leader's perspective. I am mainly focusing on analysing the pedagogy of this play episode.[9]

This vignette is situated within the vibrant research culture of the nursery school, which enables democratic practices that respect the professional identities of the teachers (Osgood, 2006) and includes children as participants in their learning (Pascal & Bertram, 2009). Pascal and Bertram emphasise the power of practitioner research as the researcher can uncover the many nuances of everyday lived experiences with children and their families. This vignette is from the same nursery as in Chapter 3, but differs since it is about children's imaginary play. It is taken from an extract of a child's learning diary, where notable observations of children's experiences in the nursery are made. They form part of the assessment process of this nursery school. This is one of many mathematical play sessions that this extremely skilled teacher has observed (Butcher, 2017; Carruthers & Butcher, 2013). She uses these observations not only to assess the children's knowledge but also to plan from them.

The nursery school is committed to children's free and imaginary play (Carruthers & Butcher, 2013). In this play episode, the children are leading, using their imagination in pretence. Here they use their own mathematical knowledge producing *their* mathematics: this is different from traditional school mathematics in England where the focus is on teaching a predetermined curriculum (DfE, 2013). In this vignette, the children make the rules and there are no right or wrong answers. They are wallowing in play ideas and, in this episode, are using their previous knowledge to try and understand the concept of keeping money in a safe. As Steffe (2004) writes, "The mathematics of the children emerges from within children and it must be constructed by children" (p. 35).

The mathematics in which the children are engaged is not traditional number knowledge of learning numbers in sequence or of standard calculations, but within the concept of safes and monies. Their numbers are not in the traditional order; you need to know the code and the children make up their own codes. This aspect seems to have fascinated the children, and in this short vignette of spontaneous pretend play, the children have used seven different combinations of the numbers, one to nine. They did this with fluency and are already meeting the expected outcomes of the EYFS (DfE, 2017), which are, for example, that children should show an interest in numerals in the environment and representing numbers; use number names and language spontaneously; and show curiosity about numbers by asking questions and offering comments.

The teacher

Emma says her pedagogy is not about "teachery teaching" but it about "being with the children." This resonates with teaching that is attached (Carruthers, 2017), where the children and their teacher are conceptually and contextually connected. There is also a calm element within Emma's description of her pedagogy, since she is not rushing to achieve a teaching objective. She has personal knowledge of the children, their family life, and their personal histories, and this makes it easier to understand their actions and thinking (Moll et al., 1992). She responds to the play spontaneously; at the same time, she understands the richness of what is happening and make notes of the play action and conversation. She is an *insider* in this scenario and a participant observer (Carruthers & Worthington, 2011). Without the written observations made at the time, reflection on this action would be difficult (Schön, 1987). Emma's acute understanding of what was important to note is vital to support the children in their continuous learning about safes and monies. For staff professional development, it is also crucial to discuss this vignette, to add to their knowledge and to our knowledge through this chapter. This observation is a nugget of enlightenment for us to dissect, reflect, rethink, and to learn from, and it is important for the teacher's professional identity that we acknowledge this contribution (Pascal & Bertram, 2015).

The teacher as researcher

This teacher has been researching children's play experiences and the link with imaginary play and mathematics (Carruthers & Butcher, 2013). The nursery school encourages and enables teachers and practitioners to pursue research interests and provides space and time for them to participate in professional conversations in what Wenger (1998) might term *a community of practice*. In a conversation with this teacher, she said that

> Play is like magic and when children get together the magic begins again. They can be anywhere at the park, on the boat. They just collaborate, and even if they have never met each other, they know they can join in, new ideas are thrown into the pot and are taken up, or not.

It is this appreciation of children's play that this teacher demonstrates through her observations, enjoying this thoughtful episode of child interaction. This is part of the pedagogy that makes play solid and useful to children (Carruthers & Worthington, 2011).

It is also about the teacher wanting to know and creating her own understanding, whilst deepening her knowledge about play. She not only uses her "on the ground" awareness of children's play, but is also aware of other play perspectives through reading relevant literature and discussing with colleagues; the combination is powerful (Carruthers, 2015). It creates new knowledge and understandings and continues to make play interesting to the teacher who can articulate her knowledge to others. For many children in English schools, play can be so structured that it becomes far removed from children's own play (Rogers, 2010): it is not encouraged or facilitated in many schools after the age of five as the children are in more formal education (Ofsted, 2017). There is a dearth of understanding about play (Rogers, 2010). Added to this, teachers and practitioners are not always confident in mathematics and are unaware of the potential of the mathematics in play and therefore may not respond to it (Markovits & Forgasz, 2017; Pound, 2008). It is unusual to find documentation of children's spontaneous pretend play sessions that include spontaneous mathematical play; however, Worthington and van Oers's research found that in a high proportion of their play observations in this nursery setting, children were engaging in a variety of aspects of mathematics.

Resources

Previous to this imaginary play episode, Emma invited mathematical happenings through resources by bringing in a safe, which she thought Isaac might be interested in. The children had already discussed and asked questions about the idea of a safe before this vignette occurred. In this present vignette, the children became more interested in the concept of a safe and focused on a critical aspect, the number code to open the safe. Isaac knows the issues around number codes as he says, "it is rather difficult to remember" this secret number.

A key feature of the pedagogy of play that this teacher shows is "her receptivity to the unpredictable" (Fochi, 2019, p. 342). Emma accommodates the pretence and adapts to the situation as she offers alternative solutions, that is, a cheque which Isaac appears not to accept by quickly saying, "I need hundreds of pounds in cash!" Emma quickly responds by finding money in her purse, eager to keep the play going. Cheques have previously been a resource that the children had found useful in their play, but the coins seemed to suffice instead of notes, and are accepted by the children as part of the imaginary play. This highlights the importance of pretence in accommodating substitutes instead of real things. This adds to the children's growing knowledge of finance and monies. The resources are therefore never random but authentic aspects of everyday mathematics of the world and acknowledge the children's threads of thinking at the time.

Availability of graphic materials

The resources of accessible graphic materials enable the children to use these tools in their play, and to communicate as they see adults communicate, in their world. It is part of their enculturation to the symbolic systems of their world (Kress, 1997). In this nursery, this graphic equipment is always available in various forms. The walls of the nursery are covered with the children's own free graphical explorations rather than the finished intricate drawings for which the Reggio Emilia nurseries in Italy are renowned; it is the emerging and experimental graphics that are displayed. Children investigate the forms of lines and other early marks (Matthews, 2003), and here they also explore their own invented and standard signs and symbols (Carruthers & Worthington, 2005).

Modelling

A pedagogical strategy that this teacher employs is naturally modelling, in authentic situations, mathematical signs in play context. For example, she wrote down the numbers Isaac requested "and looking at the numbers and symbols he took a great interest in what she had written." Humans, including children, possess an innate ability to learn by looking at and observing others (Meltzolf & Williamson, 2010). The teacher was modelling how to write numbers and their appearance in the context of an authentic purpose.

No limits on mathematics learning

Emma does not limit the children's learning but encourages them to go far as they want within the concept in which they are currently involved. She is raising expectations of mathematical knowledge, in a broad sense, as the children develop the play by widening the scope of what is in the safe to "important information." This is in sharp contrast to what Jordan (2004) has observed, which is that teachers often only stick to very low-level mathematics concepts,

> Unfortunately, in the absence for many teachers of sufficient knowledge, or interest in learning more, children are exposed to their teacher's reinforcement of lower level concepts such as colour, counting, and shape, even when such reinforcement is inappropriate or unnecessary.
>
> (p. 99)

These mathematical ideas of codes and secret numbers lasted for many weeks. Emma added a variety of safes, secret boxes, and door codes to extend the children's thinking. One of the main features of her practice is that she carefully listens to the children and takes her cues from them. She has developed an intuitive awareness of children's meaning making (Carruthers, in progress).

The teacher Emma is a *change agent* (Rodd, 2015), and I argue that it is professional colleagues like Emma who, on the ground, will change practice; they are the mediators and enablers of authentic mathematical play practices that unlock children's mathematical thinking. The pedagogy of early mathematics depicted in this vignette is not found in a list of tips for teachers, but is much deeper than this, respecting teachers' professionalism and children's intellectualism. The pedagogy of young children's mathematics can only flourish in a democratic nursery culture, with teachers who understand the context and cultures of the young children's worlds they teach; who can facilitate play opportunities, collaborating and co-constructing alternative play worlds; who have an awareness and knowledge of the breadth of mathematical learning; and who have the passion to want to delve further. This will support young children's higher-level mathematical thinking.

Commentary 3: a perspective from Denmark

Author: Lone Hattingh

Whilst early childhood education in Denmark is often held in high regard with its recognition of play and social pedagogy as a fundamental feature of education, there is nonetheless a move towards a curriculum with defined goals or themes (Broström, 2017). Brogaard Clausen (2015) makes reference to schoolification of the early childhood curriculum in England, yet the impact of evaluation and standardisation in response to international comparisons can be seen in a move to a more prescriptive curriculum in the Nordic countries, too (Jensen, 2014). Broström further states that the Danish early years curriculum has changed to include references to six themes, which include goals and developing concepts, and to preparation for school. This can be seen as part of a discourse about outcomes, performance, and the economic benefit of early childhood, with the view that children are an investment for the economic future of the country (Jensen, 2014).

However, principles of social pedagogy and democracy continue to underpin Danish preschool practice. Childcare is under the direction of the Ministry for Children and Social Affairs (2018), which publishes the curriculum for early years. The curriculum is made up of six broad learning themes and was originally introduced in Denmark in 2004 in response to criticism that there was too much free play, and that the adults working with children should have a more systematic and organised approach to their learning (Ahrenkiel, 2015). There is a focus on children's developing understanding of number and mathematical concepts in the theme for nature and science (Ministry for Children and Social Affairs, 2018), with the environment providing opportunities for developing concepts such as those of quantity, size, position, and weight. This learning theme encourages pedagogues to arrange the environment so that children have opportunities to experiment with number, area, and form in their early experiences of mathematical language and thinking. This resonates with the way in which the children's teacher, Emma, provides multimodal resources for the development of their play and mathematical thinking. Within the Danish curriculum's six learning themes, play has a strong and visible presence, and guidance is broad. This allows settings to interpret them as they consider relevant.

Klitmøller and Sommer (2015) argue that the impact of neoliberal values, whilst less visible in Denmark than in many other countries, results in separating children's learning into separate themes. Citing George Monbiot, Moss (2019, p. 16) explains that neoliberalism has become the dominant "powerful political narrative" of our time, emphasising competition, economics, and individual choice. This has brought about a discourse which is based on standardisation and outcomes. Consequently, neoliberalism has encouraged simplistic forms of evaluation and assessment which are easy to measure and therefore favoured by some governments. Neoliberalism can be seen to contradict the values of the Nordic social pedagogical approach to early years education which foreground holistic learning, education, and play (Kragh-Müller, 2017). Broström (2017) suggests that the relationship between play and learning is at times tenuous and calls for a synthesis where the "learning dimension is embedded in play" (p. 13). This embeddedness can be seen in Jayden and Isaac's play with the wooden building blocks, which they then transport to a more suitable place as they develop their imaginative play. The materials were available to the children to use spontaneously in their play: wooden building blocks, furniture (a small cupboard), keyboard, clipboard, tape, and mark-making implements. In his seminal work *Before Writing*, Kress (1997) suggests that materials need to be "to hand" so that children can select those that represent their thinking and their meanings.

The materiality of play

According to Broström (2017), it is the meaningful elements of play that facilitate children's learning. The accessibility of the materials enriched Jayden and Isaac's play and enabled them to take a multimodal approach to create a situation which reflected their own interests and their cultural understanding about the purpose, function, and usefulness of their safe. Drawing on Bishop's (1998a) "six fundamental activities," the multimodal elements of Jason and Isaac's play provide opportunities for the development of their mathematical learning in a meaningful context. In developing their thinking, Jayden and Isaac use more than language and the available materials; they also use sounds (beep, beep, beep) and actions, such as tapping on the keyboard and pressing buttons for access to the safe. The complexity of their play is evidence of their growing mathematical understanding while at the same time moving away from the binary either/or of play and learning (Broström, 2017).

Of importance for Lenz Taguchi (2010) are time and space, which are evident as the children develop their ideas and draw their teacher, Emma, into their play. In the vignette, the children's play with their safe takes place over several days, reflecting Emma's calm approach in encouraging the children to develop their ideas in their own time. Within this unhurried environment, they are able to explore their ideas in depth, showing that they are resourceful and capable of complex thinking. Pacini-Ketchabaw, Kind, and Kocher (2017, p. 9) explain how time "to slow down, to listen" provides children with opportunities to discover the "thingness" of materials around them and to develop their thinking. In this way this window onto the children's play makes visible their "intra-active relationship" (Lenz Taguchi, 2010, p. 10) with each other, their teacher, and the materials as well as the time and space which are freely available to them.

Jayden's wavy lines on the paper and clipboard represent the result of the block being taken away by another child: yet while this line may not represent numerals in the traditional sense, his understanding of its use in representing subtraction is evident in the way he responds and makes this part of his play. It is frequently the apparently unremarkable that goes unnoticed, suggesting that careful and close engagement (or entanglement, as Lenz Taguchi might suggest) with the data is needed to try to understand the development of the child's thinking. The different ways in which Isaac and Jayden use of the materials also reflect their inventiveness and resourcefulness, confirming the deep learning that takes place in their pretend play scenario. This scenario would sit comfortably within the Danish learning themes (Ministry for Children and Social Affairs, 2018), where the documentation begins with a declaration of the importance of play and a broad understanding of learning and community.

Shared voices: reflections on playing, learning and mathematics

Bishop (1988a) argues that mathematics is more than "meaningless pushing around of figures which many children still unfortunately experience" (p. 186). As this vignette shows, high-quality and sustained impromptu pretend play (Bishop's "imagined reality") can also support children's conceptual learning such as mathematics.

The authors of this chapter – from England and Denmark – value children's self-initiated and spontaneous play not as an "activity" to be *used by teachers* for teaching mathematics, but for the opportunities, experiences, and qualities it engenders in young children. In this nursery school, children draw on many modes to explore, make, and communicate their meanings, and in this, the nursery school's practice differs from much adult-led mathematical pedagogy in England. Whilst the early childhood curriculum in Denmark (Ministry for Children and Social Affairs, 2018) is made up of six themes: personal development; social development; communication and language; body, senses and movement; nature, outdoor experience and science; culture, aesthetics, and community (author's translation), the emphasis on play favours a holistic approach in which specific elements of learning are loosely defined. This contrasts to the English curriculum where mathematics appears as a distinct area within the theme for "learning and development" (DfE, 2013).

But for many in early years education, there has been an apparent division between play and learning, Broström (2017, p. 1) considering an "either/or" positioning unproductive, a view that denies a unified concept of play and learning. Such a perspective endorses a distinct divide between the two where in today's world, "learning" belongs in the teacher's domain and "play" in the child's, pretend play often narrowly circumscribed. In this vignette of Isaac and Jayden's play, their mathematical learning takes place within a context of deep cultural meaning for them, alongside much other learning, which is not so easily described as they discuss and collaborate with each other and their teacher, Emma. One might argue that the requirement to record children's learning in narrowly defined ways as required by the English *Statutory Framework for the Early Years Foundation Stage* (DfE, 2013), fails to acknowledge other learning that is taking place alongside and within this potentially rich play experience.

Broström (2017, p. 7) argues that "children learn through interaction and communication *when they experience the activity as meaningful* and when it sparks their creativity and imagination" (italics added), concluding that we can "overcome this educational conflict (play versus learning) and identify a shared understanding where the learning dimension is embedded in play" (Broström, 2017, p. 11). Thus, in reflecting on Isaac and Jayden, it is evident that their play fulfils all these criteria, suggesting that the children's experiences of play and learning, expressing the nursery school's shared democratic values and philosophy and creating something of Broström's "unified theory" of play and learning.

The children's experiences are a reflection of the high quality of teaching in this nursery school, underpinned by its philosophy and research culture, and is clearly a setting which recognises that children have a relationship with their social and material surroundings, in a "state of interdependence" (Lenz Taguchi, 2010, p. 51). With the emphasis on children's own imaginative play, the in-between spaces also have meaning. These children attend a nursery setting which encourages free and uninterrupted exploration in their play, enabling teachers to observe and assess children's learning and experiences through a wide lens without the need to rely on developmentally defined outcomes.

Whilst Denmark seems to have put play at the forefront of learning in their curriculum document, this was also true in the first early years curriculum documents in England (DfEE, 2000), which offered hope for a greater understanding of play and for making it officially recognised; but interpretations became blurred and through the years it has become more the dictate of the Office for Standards in Education (Ofsted) and government regulations (Ofsted, 2017). Like Denmark, play in this nursery school has "a strong and visible presence." The vignette of Isaac and Jayden "would sit comfortably" there, but to what extent is this always true for England?

Oberhuemer (2005) calls for "democratic professionalism" where early years teachers and practitioners need to see themselves as "interpreters and not mere implementers of curricular frameworks" (p. 13). From each of the commentaries it was clear that the teacher in the vignette characterised democratic professionalism;

but teachers like Emma are perhaps atypical, at least, in England. Perhaps, therefore, all professional development for early childhood teachers and practitioners needs to consider emphasising professional identities, critical reflection, and teachers as agents of their own learning (Rodd, 2015).

Notes

1 Nursery schools in England are a unique early years' provision that are government funded. They have a tradition dating back to the 1920s, rooted in the work of Margaret Macmillan and Susan Isaacs (Giardiello, 2013).
2 Both teachers and early years' practitioners work in the nursery school, but for brevity, the word "teacher" is used throughout.
3 Regrettably, Isaac's mathematical inscription is not sufficiently clear to reproduce it here.
4 The nursery's approach to mathematics developed from work by Carruthers and Worthington.
5 Prior to four years of age, some children will attend nursery schools, whilst others may attend preschool playgroups or childminders.
6 The teaching of "synthetic phonics" especially, has had a significant impact on all aspects of teaching and learning in the Foundation Stage in England. In turn, this has exerted new pressures on early years mathematics.
7 The four rules listed here and the explanation of each are taken from van Oers (2013, pp. 191–192).
   In this instance, technology appears to be a dominant interest (see, for example, Pirani and Hussain (2019), who explore the value of technology for learning in the early years).
8 The term "hot spot" relates to biodiversity and describes a region that is a significant reservoir of biodiverse species. The diversity (of children's mathematical thinking, talk, conceptual knowledge, and graphical representations) is exemplified in this and other chapters, thereby endorsing it as a *mathematical* (and play) hot spot.
9 The title "National Teaching School" is awarded to a school which has previously developed research and is a centre of initial teacher training: these schools are funded for their work.

References

Ahrenkiel, A. (2015). Pædagogfagligheden under pres: Mod målstyring af pædagogikken [Pedagogical professionalism under pressure: Against management by objectives in education]. In J. Klitmøller & D. Sommer (Eds.), *Læring, dannelse og udvikling* [Learning, education and development], (pp. 41–59). Copenhagen: Hans Reitzels Forlag.

Ball, J., & Olmedo, A. (2013) Care of the self, resistance and subjectivity under neoliberal governmentalities, *Critical Studies in Education, 54*(1), 85–96. https://doi.org/10.1080/17508487.2013.740678.

Bishop, A. J. (1988a). Mathematics education in its cultural contexts. *Educational Studies in Mathematics, 19*(2), 179–191. Retrieved from https://link.springer.com/article/10.1007/BF00751231.

Bishop, A. J. (1988b). The interactions of mathematics education with culture. *Cultural Dynamics, 1*(145), 145–157. https://doi.org/10.1177/092137408800100202.

Bishop, A. J. (1991). *Mathematical enculturation: A cultural perspective on mathematics education,* (2nd ed.). Dordrecht: Kluwer.

Brogaard Clausen, S. (2015). Schoolification or early years democracy? A cross-curricular perspective from Denmark and England. *Contemporary Issues in Early Childhood, 16*(4), 355–373. https://doi.org/10.1177/1463949115616327.

Broström, S. (2017). A dynamic learning concept in early years education: A possible way to prevent schoolification. *International Journal of Early Years Education, 25*(1), 3–15. https://doi.org/10.1080/09669760.2016.1270196.

Butcher, E. (2017). The mathematical intentions and encounters of 2- and 3-year olds. *Early Educational Journal, 83*(Special Issue: Early Years Mathematics), 7–9.

Carruthers, E. (2015). Listening to children's mathematics in school. In B. Perry, A. Gervasoni, & A. Macdonald (Eds.), *Mathematics and transition to school: International perspectives*, (pp. 313–330). Sydney: Springer.

Carruthers, E. (2017). Children's mathematics. *Early Education Journal, 83*(Special Issue: Early Years Mathematics), 4–6.

Carruthers, E. (in progress). *The pedagogy of children's mathematical graphics in calculation.* [PhD dissertation, Bristol: Bristol University].

Carruthers, E., & Butcher, E. (2013). Mathematics: Young children co-construct their mathematical enquiries. In P. Beckley (Ed.), *The new early years foundation stage: Changes, challenges and reflections* (pp. 91–104). Maidenhead: Open University.

Carruthers, E., & Worthington, M. (2005). Making sense of mathematical graphics: The development of understanding abstract symbolism. *European Early Childhood Education Research Journal, 13*(1), 57–79. https://doi.org/10.1007/s10649-019-09898-3.

Carruthers, E., & Worthington, M. (2006). *Children's mathematics: Making marks, making meaning.* London: SAGE.

Carruthers, E., & Worthington, M. (2011). *Understanding children's mathematical graphics: Beginnings in play.* Maidenhead: Open University.

Cook, D. (2006). Mathematical sense making and role play in the nursery. *Early Child Development and Care, 121*(1), 55–66. https://doi-org.vu-nl.idm.oclc.org/10.1080/0300443961210106.

Dahlberg, G., Moss, P., & Pence, A. (2007). *Beyond quality in early childhood education and care,* (2nd ed.). London: Routledge.

Danniels, E., & Pyle, A. (2018). Defining play-based learning. OISE University of Toronto, Canada. Retrieved from www.child-encyclopedia.com/sites/default/files/textes-experts/en/4978/defining-play-based-learning.pdf.

DfE (2013). *The Primary National Curriculum in England: Key Stages 1 & 2 framework document.* London: Department for Education. Retrieved from https://assets.publishing.service.gov.uk/government/uploads/system/uploads/attachment_data/file/425601/PRIMARY_national_curriculum.pdf

DfE (2017). *Statutory framework for the Early Years Foundation Stage.* London: Department for Education. Retrieved from www.foundationyears.org.uk/files/2017/03/EYFS_STATUTORY_FAMEWORK_2017.pdf.

DfEE (2000). *Curriculum guidance for the foundation stage.* London: Qualifications and Curriculum Authority.

Fochi, P. (2019). Pedagogical documentation as a strategy to develop praxeological knowledge: the case of the observatory of childhood culture – OBECI. *European Early Childhood Education Research Journal, 27*(3), 334–345. https://doi.org/10.1080/1350293X.2019.16008.

Giardiello, P. (2013). *Pioneers in early childhood education: The roots and legacies of Rachel and Margaret McMillan, Maria Montessori and Susan Isaacs.* Abingdon: Routledge.

Gifford, S. (2005). *Teaching mathematics 3–5.* Maidenhead: Open University.

Hewes, J. (2014). Seeking balance in motion: the role of spontaneous free play in promoting social and emotional health in early child care and education. *Children, 1,* 280–301. https://doi.org/10.3390/children1030280.

Jensen, A. S. (2014). The deluge. *European Early Childhood Education Research Journal, 22*(1), 77–90. https://doi.org/10.1080/1350293X.2013.865358.

Jordan, B. (2004). Scaffolding learning and co-constructing understandings. In A. Anning, J. Cullen., & M. Fleer (Eds.), *Early childhood education: Society and culture,* (pp. 31–42). London: SAGE.

Klitmøller., J., & Sommer, D. (2015). Børn i institution og skole: Læring, dannelse og udvikling [Children in day care and school: Learning, education and development]. In J. Klitmøller. & D. Sommer (Eds.), *Læring, dannelse og udvikling* [Learning, education and development], (pp. 9–38). Copenhagen: Hans Reitzels Forlag.

Kragh-Müller, G. (2017). The key characteristics of Danish/Nordic child care and culture. In C. Ringsmose & G. Kragh-Müller (Eds.), *Nordic social pedagogical approach to early years,* (pp. 3–23). Switzerland: Springer.

Kress, G. (1997). *Before writing: Rethinking the paths to literacy.* London: Routledge.

Lenz Taguchi, H. (2010). *Going beyond the theory/practice divide in early childhood education.* London: Routledge.

Markovits, Z., & Forgasz, H. (2017). Prospective preschool teachers beliefs before and after a mathematics teaching course. *Education Alternatives Journal of International Scientific Publications, 15,* 80–89. Retrieved from www.scientific-publications.net/en/article/1001564/

Matthews, J. (2003). *Drawing and painting: Children and visual representations.* London: Paul Chapman.

Meltzolf, A. N., & Williamson, R. A. (2010). The importance of imitation for theories of social-cognitive development. In G. Bremer, & T. Wachs (Eds.), *Handbook of infant development* (2nd ed.) (pp. 345–364). Oxford, England: Wiley Blackwell.

Ministry for Children and Social Affairs. (2018). *Den styrkede pædagogiske læreplan, Rammer og indhold* [The reinforced educational curriculum, Frameworks and content]. Copenhagen: Ministry for children and social affairs. Retrieved from https://arkiv.emu. dk/sites/default/files/7044%20EVA%20SPL%20Publikation_web.pdf

Moll, L., Amanti, C., Neff, D., & Gonzales, N. (1992). Funds of knowledge for teaching. *Theory into practice, 31*(2), 132–141. https://doi.org/10.1080/00405849209543534.

Moss, P. (Ed.). (2013). *Early childhood and compulsory education: Reconceptualising the relationship.* London: Routledge.

Moss, P. (2019). *Alternative narratives in early childhood: An introduction for students and practitioners.* London: Routledge.

Munn, P., & Kleinberg, S. (2003). Describing good practice in the early years – a response to the 'third way'. *Education 3–13, 31*(2), 50–53. https://doi.org/10.1080/03004270385200221.

Munn, P., & Schaffer, R. (1993). Literacy and numeracy events in social interactive contexts. *International Journal of Early Years Education, I*(3), 81–80.

Oberhuemer, P. (2005). Conceptualising the early childhood pedagogue: Policy approaches and issues of professionalism. *European Early Childhood Education Research Journal, 13*(1), 5–16. https://doi.org/10.1080/13502930585209521.

Ofsted. (2017). *Bold beginnings: The reception curriculum in a sample of good and outstanding primary schools* [Review]. Retrieved from https://schoolsweek.co.uk/wp-content/uploads/2017/11/28933-Ofsted-Early-Years-Curriculum-Report.pdf.

Osgood, J. (2006). Deconstructing professionalism in early childhood education: Resisting the regulatory gaze. *Contemporary Issues in Early Childhood, 7*(1), 5–14. https://doi.org/10.2304/ciec.2006.7.1.5.

Pacini-Ketchabaw, V., Kind, S., & Kocher, L. (2017). *Encounters with materials in early childhood education*. London, England: Routledge.

Pascal, C., & Bertram, A. (2009). Listening to young citizens: The struggle to make real a participatory paradigm in research with young children. *European Early Childhood Education Research Journal, 17*(2), 249–262. https://doi.org/10.1080/13502930902951486.

Pascal, C., & Bertram, T. (2015). *Praxis, ethics and power: Developing praxeology as a participatory paradigm for early childhood research*. Birmingham: Centre for Research in Early Childhood.

Pirani, S., & Hussain, N. (2019.) Technology is a tool for learning: Voices of teachers and parents of young children. *Journal of Education & Social Sciences, 7*(1): 55–66. https://doi.org/10.20547/jess0711907105.

Pound, L. (2008). *Supporting mathematical development in the early years*. Maidenhead: Open University.

Roberts-Holmes, G. (2015). The "datafication" of early years pedagogy: "If the teaching is good, the data should be good and if there's bad teaching, there is bad data." *Journal of Education Policy, 30*(3), 1–13. https://doi.org/10.1080/02680939.2014.924561.

Rodd, J. (2015). *Leading change in the early years: Principles and practice*. Maidenhead: Open University.

Rogers, S. (2010). Powerful pedagogies and playful resistance: Role play in the early childhood classroom. In L. Brooker. & S. Edwards (Eds.), *Engaging play* (pp. 152–165). Maidenhead: Open University.

Schön, D. A. (1987). *Educating the reflective practitioner*. San Francisco, CA: Jossey Bass.

Seo, K-H., & Ginsburg, H. (2004). What is developmentally appropriate in early childhood mathematics education? Lessons from new research. In D. Clements., J. Sarama., & A. M. DiBaise (Eds.), *Engaging young children in mathematics: Standards for early childhood mathematics education* (p. 91–104). Mahwah: Lawrence Erlbaum.

Steffe, L. (2004). PSSM (Principles and standards for school mathematics) from a constructivist perspective. In D. H. Clements & J. Sarama (Eds.), *Engaging young children in mathematics*, (pp. 26–34). London: Lawrence Erlbaum.

van Oers, B. (2010). Emergent mathematical thinking in the context of play. *Educational Studies in Mathematics, 74*(1), 23–37. https://doi.org/10.1007/s10649-009-9225-x

van Oers, B. (2013). Is it play? Towards a reconceptualisation of role play from an activity theory perspective. *European Early Childhood Education Research Journal, 21*(2), 185–198. http://dx.doi.org/10.1080/1350293X.2013.789199.

Vygotsky, L. S. (1978). *Mind in society: The development of higher psychological processes*. Cambridge, MA: Harvard University Press.

Weisberg, D. S., Kittredge, A. K., Hirsh-Pasek, K., Golinkoff, R. M., & Klahr, D. (2015). Making play work for education, *Phi Delta Kappan, 96*(8), 8–13. https://doi.org/10.1177%2F0031721715583955.

Wenger, E. (1998). *Communities of practice: Learning, meaning, and identity*. Cambridge, England: Cambridge University.

Worthington, M. (2018). Funds of knowledge: Children's cultural ways of knowing mathematics. In M-Y Lai, T. Muir, & V. Kinnear (Eds.). *Forging connections in early mathematics teaching and learning* (pp. 239–258). Singapore: Springer.

Worthington, M. (2020). Children playing pretend and the cultural transmission of mathematical signs. In A. MacDonald, L. Danaia, & S. Murphy (Eds.), *STEM education across the learning continuum: Early childhood to senior secondary* (in press). Singapore: Springer.

Worthington, M., & van Oers, B. (2016). Pretend play and the cultural foundations of mathematics. *European Early Childhood Education Research Journal, 24*(1), 51–66. https://doi.org/10.1080/1350293X.2015.1120520.

PART 7

Conclusion

12

REAPING THE BENEFITS OF REFLEXIVE RESEARCH AND PRACTICE IN EARLY CHILDHOOD MATHEMATICS EDUCATION

Continuing the conversation

Oliver Thiel, Elena Severina, and Bob Perry

Innovative pedagogy

The title of this book is *Mathematics in Early Childhood: Research, Reflexive Practice and Innovative Pedagogy*. In this final chapter, we will focus on these three aspects,

- a discussion of what we mean by *innovative pedagogy*
- a deeper analysis of how *reflexive practice* works
- recommendations for *future research* in the field

One of the great achievements of Alan Bishop's (1988a, 1988b) research is that he revealed the relationship between mathematics and culture. Since the ancient Greek philosopher Plato, most mathematicians have believed that abstract objects (like numbers) and the mathematical relations between them exist independently from the human mind (Marcus, 2015). This has to be true if Galileo Galilei (1623) is right when he says that the universe "is written in mathematical language, and the letters are triangles, circles and other geometrical figures" (p. 25). Bishop confirms this when he identifies six *universal* mathematical activities that he found exist across the world despite cultural differences. Even though the mathematical objects, relations, and activities are universal, the way in which humans represent mathematical objects, express mathematical relations, and work with mathematical activities can be very different (Bishop, 1988a). This is particularly so when we focus on how humans pass mathematical knowledge on to the next generation, that is, when we consider mathematics education (Bishop, 2002).

Because of the cultural differences in education, it is challenging to write a book about *innovative pedagogy*. When reading the chapters of this book, you might have had the impression that not all approaches are equally innovative. Some chapters may describe something that is common practice in your culture. What you consider ordinary or extraordinary can be different for each of you, and Bishop's theory can help us understand the differences.

I, Oliver Thiel, met Alan Bishop at the 13th International Congress on Mathematical Education (ICME) in 2016 in Hamburg. He was pleased that the research that I presented at the conference (Nakken et al., 2016) was related to his work and that we used *all six* of his activities. He remarked that it is widely agreed that mathematics is about numbers (*counting*), measurement (*measuring*), shapes (*designing*), and space (*locating*), but he often experienced less emphasis on *explaining* and *playing*. Whether one places more emphasis on explaining or on playing is related to one's cultural background, especially the pedagogical tradition.

In Europe, there are two different traditions of early childhood education and care that we want to exemplify by comparing the UK and Norway. The UK has an *early education* or *readiness for school tradition* focusing "strongly on cognitive development, early literacy and numeracy" (OECD, 2006, p. 136). This can be seen in the prescribed ministerial curriculum (DfE, 2017) with detailed learning goals. Norway follows the *social pedagogy* or *Nordic tradition*. It has no curriculum for the early years, only a Framework Plan (Ministry of Education and Research, 2017) outlining values and requirements of kindergarten education. The Framework Plan does not address what children should learn. It provides only "guidelines for local authorities and the centres about the values, purposes and processes of early childhood education and care" (OECD, 2006, p. 138). There is a strong tradition of free play that should not be disturbed by adults. To talk about "teaching" in kindergarten is a taboo (Sæbbe & Pramling Samuelsson, 2017).

Looking at the chapters of this book that originate from the UK (Chapters 3 and 11) and Norway (Chapters 2, 7, and 8) reveals an interesting pattern. In both countries, the authors somehow write against their tradition. The chapters from the UK place a strong emphasis on playing, and as Alf Coles (in Commentary 1 of Chapter 3) highlights, the teacher "is not seeking for the children to explain what they are doing." Early years teachers in the UK need to read this. For them, it is innovative to facilitate child-centred, child-directed free play that is not aimed at predefined learning goals. But that is not what Norwegian kindergarten teachers need to read. In Norway, free play in kindergarten is common practice. Karlsen and Lekhal (2019) found that children in Norwegian kindergartens spent about 60% of the day in free play. The other 40% "consisted of meals, circle time, quiet time, and gym, all of which primarily involved routines rather than learning-centred activities" (Karlsen and Lekhal, 2019, pp. 238–239). In free play, they observed that practitioners often were absent or showed "limited evidence of supportive interactions (measured as joining in, commenting, helping and instructing) between practitioners and children during free play when practitioners were present" (Karlsen and Lekhal, 2019, p. 242). Against this tradition, the chapters from Norway show teachers who support and enrich children's play, ask children to explain their mathematical ideas, and facilitate children's mathematical thinking in playful ways. This is innovative pedagogy in Norway, but it carries an inherent danger of schoolification – a topic that is controversially discussed in Norway (Tuastad, Bjørnestad, & Alvestad 2020) and other countries.

Comparing all chapters, we find different examples of child-initiated and teacher-initiated activities that are guided, directed, and governed in different degrees by the teacher or the children. Each example is innovative in its own cultural context, but it might be inspirational for readers from other cultural backgrounds, too. Bishop has shown that all six activities are equally important despite cultural differences. As a whole, the chapters show the richness of early mathematics and provide early childhood educators with many ideas on how to create learning opportunities for children to explore and experience all six of Bishop's fundamental mathematical activities by explaining and playing with counting, measuring, locating, and designing.

Reflexive practice

According to Einstein (1954), "Pure logical thinking cannot yield us any knowledge of the empirical world; all knowledge of reality starts from experience and ends in it" (p. 271). It may surprise to hear this from a theoretical physicist who is best known for his abstract theories that seem to contradict our daily experiences. However, when Einstein developed his theory, he was working as a clerk in the Swiss patent office in Bern (White & Gribbin, 1994). In this position, he was confronted with the technical challenges of this era including how to synchronise clocks over large distances in order to coordinate train traffic in the fast-growing European railway network. Trains and clocks play an important role in Einstein's thought experiments that deal with the issue of simultaneity and finally led him to the counterintuitive ideas of special relativity (White & Gribbin, 1994). This matches the viewpoint of John Dewey (1916), who argued that "if knowledge comes from the impressions made upon us by natural objects, it is impossible to procure knowledge without the use of objects which impress the mind" (p. 313). His pedagogical theory of "having an experience" (Dewey, 1934, p. 35) emphasises the importance of experience and reflection. Reflection "emancipates us from merely impulsive and merely routine activity" and it "gives an individual an increased power of control" (Dewey, 1933, p. 17 and 21). Eating, for example, is a routine activity, and often young children are just given a plate that was filled by an adult. Reflecting on potato sizes and numbers (Chapter 2) gives children the skills to decide by themselves how many and what size potatoes they want to eat.

This theory was further developed by Dewey's colleague Kolb (2015) into the *Experiential Learning Theory* (ELT). Kolb describes learning as a cycle with four steps: (CE) *Concrete Experience* – having an experience while doing something; (RO) *Reflective Observation* – reviewing what you have done and reflecting on the experience; (AC) *Abstract Conceptualisation* – concluding and learning from the experience; and (AE) *Active Experimentation* – planning and trying out what you have learned, which leads to a new concrete experience. This model provides a framework for reflection and will lead to reflexivity. As described in Chapter 1, reflexivity starts with asking, "What happened?" at the RO step, continues with analysing "Why did it happen?" at the AC step, and leads to actions that will change our own behaviour at the AE step.

The Experiential Learning Cycle

Chapter 1 urges early childhood educators to move towards reflexivity, "to consider what they and others believe and value, and to undertake their practice on the basis of such reflexivity." We believe that a version of the Experiential Learning Theory proposed by Connors and Seifer (2005) can be a useful tool in this process. This variant has five steps and is combined with questions by Borton (1970). As with Kolb's cycle, it starts with (1) *experiencing*. This is the activity phase when children are engaged in one or more of Bishop's six fundamental mathematical activities. The next step is (2) *sharing*. If you want to exchange your impressions, ideas, and observations with others, you have to ask yourself, "*What* happened?" Afterwards, you can discuss together if you find any patterns. This (3) *processing* will help you answer the question "*So what* does this mean for me?" The meaning making helps (4) *generalising* from the special case that you experienced to principles that you can apply to a wider range of situations. The last step is about planning how you can use what you have learnt by asking, "*Now, what* can I do?" and (5) *applying* it. This leads to new experiencing and closes the cycle. Figure 12.1 shows all these components. In the following paragraphs, we describe the steps in more detail and apply the theory to both young children's mathematics learning as it is presented in the previous chapters and early childhood educators' professional development.

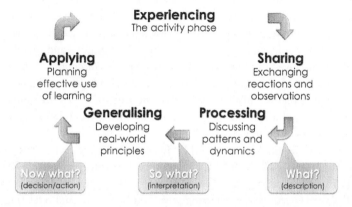

FIGURE 12.1 An experiential learning cycle with five steps (adapted from Connors & Seifer, 2005)

Experiencing: the activity phase

Kolb and Kolb (2018) complain that in many sources you can find on the internet, the activity phase of the experiential learning cycle is misunderstood. Experiencing is not merely doing. Many of our everyday activities and experiences are routine, automatic, and habitual, and do not change our behaviour, beliefs, or understanding. Here-and-now experiencing, however, can initiate learning when it "violates the expectations of previous convictions and habits of thought" (Kolb & Kolb, 2018, p. 9).

The way of experiencing mathematics is culture-dependent, while the fundamental mathematical activities (Bishop, 1988a, b) are cross-cultural. Learning gained from experiences depends, among others, on the affordances of the investigated objects and features highlighted through communication. The vignettes presented in this book, exemplify children experiencing several mathematical ideas across all six fundamental activities.

- *Explaining.* Children in Turkey classify objects by their shape, explaining for each object why it belongs to that class (Chapter 10). While harvesting potatoes in Norway (Chapter 2), children sort the potatoes by size, explaining that there cannot be two smallest ones. Other Norwegian children experience logical thinking and argumentation while explaining the intended positioning of the images in a photobook (Chapter 8). While exploring different routes through a fox burrow (Chapter 7), children solve the problem of how to avoid counting the same route twice, and explain the solution to each other.
- *Counting.* In Chapter 3, Ben from the U.K. experiences how to write large numerals. Some German children experience numbers by counting wooden bricks (Chapter 4), while Norwegian children experience the quantity of a larger set when counting how many potatoes they harvest (Chapter 2), and Isaac and Jayden from the U.K. experience numbers by playing with a pretend safe (Chapter 11).
- *Measuring.* Harry from Australia experiences that he can measure the length of a ladder by counting rungs (Chapter 5); in the potato harvest children used a "large child long" unit (Chapter 2); and in Portugal children have experiences with a measuring tape (Chapter 6).
- *Locating.* Experiencing orientation and location is evident when Julia, Samuel, and Norah are creating and exploring the routes in a fox burrow (Chapter 7), and when another group of Norwegian children are designing a photobook (Chapter 8). The Turkish children also experience locating, but within the digital environment of a game (Chapter 10).
- *Designing.* Experiencing geometrical shapes happens through music and sounds in Spain (Chapter 9), while some Turkish children engage with a digital game (Chapter 10).
- *Playing.* We find examples of children having experiences with mathematics in play in almost every chapter. For example, children play with big numerals

(Chapter 3); Harry plays with a ruler (Chapter 5); children play with a measuring tape (Chapter 6); children pretend to be a fox who seeks a way out of its burrow (Chapter 7); children play with sounds (Chapter 9); and Jayden and Isaac play with a pretend safe (Chapter 11).

While the children experience mathematical structures, their educators have the chance to observe and experience how children learn mathematics. Those experiences are not merely emotional or practical, but through reflection, they can lead to a deeper theoretical understanding of mathematics, pedagogy, and children's development. Schön (1983) emphasises that professionals not only reflect *on* their actions after an experience, but as well while they are engaged *in* an experience. He calls these different types *reflection-on-action* and *reflection-in-action*. Reflection-*in*-action means to stop in the midst of the action, in order to make adjustments and change methods, if this is necessary to improve practice. It is immediately significant for the action (Schön, 1987). We can see this in Chapter 7 when Emma recognises the children's rising tension and frustration because of the collapsing burrow. She immediately reflects on the problem and suggests a solution. She offers an idea to strengthen the construction in order to support the continuance of the children's play.

Sharing: exchanging reactions and observations

Kolb (2015) does not explicitly mention sharing in his Experiential Learning Cycle because many learning situations start with shared direct experience that "brings the subject into the room" and democratises the learning process between the educator and the learners (Kolb & Kolb, 2018, p. 12). Sharing observations, ideas, and discoveries is important for children's learning. Most of the vignettes in this book are about groups of children who have shared experiences. In some of the vignettes, we see how early childhood educators encourage children to share their experiences. In Chapter 3, the educator repeatedly invites the children to pay attention to what the other children are doing, "Come and see what Ben's doing," "You see that, Ash? Look what Aarav's doing," "Look at this, Rashad!," and so on. In some vignettes, a child shares his or her experience with an adult: in Chapter 5, Harry explains to Sonya what he did, "It's Jack's beanstalk ladder, so high, high, whoa …"

When educators are exchanging observations and experiences, they have to reflect on what happened and how they think the children perceived the event and them (Bolton, 2010). Following the developmental framework by Borton (1970), you should ask yourself "What?" questions. What happened? What did I do? What did the children do? What was I trying to achieve? What was good or bad about the experiences for the children and for me? Borton (1970, p. 88) calls this "Sensing out the differences between response, actual effect, and intended effect."

Processing: discussing patterns and dynamics

In everyday life, we share our experiences because it is delightful, as the proverb states: "A joy shared, is a joy doubled." In the context of experiential learning, the purpose of sharing experiences is to discuss them with our peers in order to reflect on them together. This reflection is essential for mathematical learning. Nakken and Thiel (2019, p. 26) argue with reference to Dewey (1934) that children have to reflect on their actions in order to develop mathematical insight. It is not enough to just *do* something. Only when we reflect on our actions and observations, are we able to discover patterns – and this is what mathematics, the science of patterns (Devlin, 1994; Steen, 1988), is about.

It is the educator's task to stimulate children to wonder, to reflect, and to search for patterns. This can be done by asking open-ended questions, that is, a question that cannot be answered by just recalling what you already know. In Chapter 2, the teacher asks if it is possible that there are many smallest potatoes, or just one. The Norwegian language distinguishes between singular *den minste* (the one that is smallest) and plural *de minste* (the more than one that are smallest). The question stimulates the children to investigate and reflect on the problem of putting potatoes in order by size when it is difficult to see the differences. Finally, the children have an "experience of thought" (Dewey, 1934, p. 37), that is, an experience that does not involve physical action and perception, but that happens when we reach or draw a conclusion. In the given example, the children's conclusion was that "smallest" was not a practical criterion to single out a winner potato. They chose the cutest potato instead. It is not mentioned how they figured this out – perhaps by a poll.

Experiencing is not limited to the activity phase. It can happen in every part of the learning cycle (Kolb & Kolb, 2018). When exchanging reactions and observations, you can experience that others had similar experiences as you had. While discussing patterns and dynamics, you can have an aha experience, that is, a sudden insight when you discover an underlying pattern. Reaching a conclusion when generalising the pattern is an experience of thought. Finally, when applying your insight to a new situation, you will experience the consequences.

Borton (1970) illustrates the processing with the question "So what?," which is about "transforming the information into immediately relevant patterns of meaning" (p. 88). When reflecting on the chapters of this book, educators can use the reflective questions given in each chapter. Here are some examples.

- *So what* characteristics of the potatoes and the activity do you think were important to provoke the children's exploration and thinking? (Chapter 2)
- *So what* mathematical concepts are the children individually working with? (Chapter 3)
- *So what* do you think the role of the ruler was in the play situation? (Chapter 5)
- *So what* skills does Emma use in order to be able to change the offers she makes, in light of feedback from the children? (Chapter 7)

- *So in what* ways might children's embodied actions contribute to their mathematical understanding? (Chapter 8)
- *So what* do you think about the ways in which digital technology is used in the learning event depicted in the vignette? (Chapter 10)
- *So what* skills, knowledge, and understandings do teachers need in order to be able to effectively identify and respond to opportunities to support mathematical thinking and learning in spontaneous play situations? (Chapter 11)

Generalisation: developing real-world principles

The main purpose of searching for patterns is to develop a general understanding. Non-mathematicians are sometimes frightened by mathematics because it is so abstract (Devlin, 2012), but with the right understanding of *abstraction,* it becomes a powerful tool that is accessible to everyone. *Abstract* means independent of our concrete experiences here and now (Donaldson, 1978; van Oers, 2001). The learner discovers a mathematical pattern in a concrete situation, but it is not limited to this special case. Generalisation means to understand that the same pattern can be applied to many different situations and contexts, including every science, technology, engineering, philosophy, and many different everyday situations.

Comparing potatoes by size (Chapter 2) is not ultimately about potatoes. The experience and reflection on it will help children develop a general concept of size that can be applied to other situations, such as the size of beanstalks and ladders (Chapter 5) and children (Chapter 6). Counting pips on dice and comparing the number to a number of bricks (Chapter 4) is not only about dice and bricks. It will help the children develop a general concept of number that can be applied to other situations, such as counting potatoes (Chapter 2) or money (Chapter 11). Exploring a fox burrow (Chapter 7) is, of course, about foxes, but in addition, the experience and reflection on it will help children develop a general concept of space that can be applied to other situations, such as stairs and trees in the forest (Chapter 8).

Abstract thinking is powerful, but it is difficult to learn – however, perhaps not as difficult as you might think. While Piaget (1936) thought that children were not capable of abstract logical thinking before they reached the formal operational stage at about seven years of age, newer research shows that this starts much earlier (Clements & Sarama, 2009; Fuson, 1992; Sarama & Clements, 2009). The gap between concrete and abstract thought can be bridged by play (Poland & van Oers, 2007). Otsuka and Jay (2017) found three features that support children's transition from concrete to abstract thinking during block play.

1   *The child shares his or her thinking with other children or adults.* This is directly linked to step two of the experiential learning cycle (see Figure 12.1).
2   *There is enough time for the child to pause for reflection.* The experiential learning cycle, especially step three, encourages reflection. In addition, there should be time for reflection-*in*-action in every part of the learning process.

3   *The child displays satisfaction over what he or she has achieved by a self-directed activity.*
In the vignettes in this book, there are examples of satisfaction, even though
young children do not necessarily express their satisfaction with words. In
Chapter 8, for example, it is mentioned several times that the children laughed
after they had solved a problem and were satisfied with the result. In Chapter 10,
Ela cries, "yeees" because she is excited about her discovery. For Dewey (1934),
satisfaction is one of the key features that distinguish the continuous experiencing of events from having an experience that has consequences for our future
behaviour.

> In contrast with such experience, we have *an* experience when the material
> experienced runs its course to fulfilment. Then and then only is it integrated
> within and demarcated in the general stream of experience from other
> experiences. A piece of work is finished in a way that is satisfactory; a problem
> receives its solution; a game is played through; a situation, whether that of
> eating a meal, playing a game of chess, carrying on a conversation, writing a
> book, or taking part in a political campaign, is so rounded out that its close
> is a consummation and not a cessation. Such an experience is a whole and
> carries with it its own individualizing quality and self-sufficiency. It is *an*
> experience.
>
> *(Dewey, 1934, pp. 36–37, italics in original)*

This fulfilment is nicely expressed in Chapter 2 when the "potatoes finally end
as a shared meal prepared by the children and staff together." Even though it is
important that a learning activity is finished in a way that is fulfilling and satisfac-
tory for the children, the learning never stops. Every answer raises new questions
and every new skill is ready to be used. The learning cycle continues.

For the reflexive educator, the step from generalisation and abstract conceptualisation to applying the discovered principles is supported by Borton's "Now what?"
questions. While the question "So what?" focuses on understanding the pattern,
"Now what?" is about future implications and applications. When reflecting on the
chapters of this book, educators can use the following reflective questions given in
each chapter as examples.

- *Now what* other experiences in outdoor environments could generate
  interactions among children and adults? (Chapter 2)
- *Now what* are possible ways to promote experiential learning in early childhood
  with everyday objects, that is, blocks, dice, pencils, etc. (Chapter 4)
- *Now what* is the potential of the situation for the development of number
  sense? (Chapter 5)
- *Now what* other everyday objects would you consider useful (to mathematics
  learning) to place in the children's doll's house? (Chapter 6)
- *Now what* could you do to take this activity forward? (Chapter 9)

- *Now what* can you do to avoid that young children experience standard geometrical shapes in stereotypical ways, for example, with no rotations or reversals? (Chapter 10)
- *Now what* are the possibilities to support opportunity for mathematical learning through play, particularly in settings where there is a "push-down" approach from school? (Chapter 11)

Applying: planning effective use of learning

The aim of generalisation is to enable the learner to handle new situations. That is how van Hiele (1957; 1986) defines *insight*: "Insight exists when a person acts in a new situation adequately and with intention" (p. 24). The person does not act at random, but according to a pattern, a mental structure that represents an abstract conceptualisation that can be applied to many different situations and contexts.

Kolb (2015) called this step of his Experiential Learning Theory *Active Experimentation* because there is no distinct way from the experience to the application. Often, reflection on experience leads to a more or less vague understanding of the underlying pattern and some possible implications. Then, the learners have to try out if the interpretation was suitable by doing an experiment of planning and applying what they have learned to a new situation. This will happen after the activities that are described in the vignettes. Possibilities for those applications are mentioned in the commentaries. For example, can the children apply the spatial concepts that they have developed during the photobook activity (Chapter 8) when they try to create a map by using pictures taken in their neighbourhood? Can Harry apply the concept of measuring that he developed by making Jack's beanstalk ladder (Chapter 5) when he explores the growth of real beanstalks?

For the reflexive educator, this step of the Experiential Learning Cycle is about actively changing your behaviour according to your answers to the "Now what?" questions above. This can be challenging, and it might be helpful for educators to find "critical friends" (Bambino, 2002). Critical friends are colleagues who meet on a regular basis to look at each other's practice. They use outside sources and their own experience to find out what constitutes good teaching and learning. They visit each other in the classroom, give feedback, and share what works best for children's learning (Cushman, 1998).

Applications for early childhood educators

The editors and authors of this book hope that it will help you to kick-start or extend your professional development in early childhood mathematics education. Both the vignettes and the commentaries might have given you new ideas and a broader perspective on mathematics in early childhood. Is there something that caught your interest? Something that talks to you and fires your imagination? Reflect on how you can adapt it to your conditions, apply it in your context, and start doing it. Remember that the application is an experiment. It does not matter if the result of the first try-out is not what you expected. Regardless of the out-come, you will have concrete experiences that can start up a cycle of experiencing, sharing observations, processing and discussing patterns, generalising, and applying what is learnt. This will lead to new experiences – and as you go along, you will develop a deeper and deeper understanding and more and more advanced skills. Thus, the cycle is a spiral rather than a circle. Each completion of the cycle leads to new experiences "with new insight gained by reflection, thought, and action. Thus, the experiential learning spiral describes how learning from experience leads to development" (Kolb & Kolb, 2012, p. 1212).

Applications for teacher educators

The editors and authors of the book hope that it helps you and your students be more reflexive in your practice. This book is firmly centred on commentary and critique of existing practice in a variety of contexts, resulting in demonstrations of reflexive practice and research similar to those espoused as exemplary approaches in the mathematics education practice of early childhood educators and researchers. Each chapter is a result of the authors' Learning Cycle, but it also can be seen as experiencing one of your own. Was there any chapter that particularly drew your attention and made you stop and think? Perhaps it was something unexpected. Teacher educators might use a chapter of the book for professional development by asking questions like "What happened?," "Why did it happen?," "How different would it be in my cultural context?," and "What implications for my teaching prac-tice could it have?" Another opportunity we can see is to use the vignettes from the book in your practice to exemplify Bishop's mathematical activities across coun-tries. The vignettes might also be used in group discussions about, for example, the diversity of the pedagogical practice or specific mathematical ideas. The commen-taries could be then used to exemplify what might be different from a researcher's and a practitioner's point of view. For teacher educators, the reflexive questions might be very useful in stimulating discussion, providing rich assessment tasks, and synthesising data derived from consideration of the vignettes. All of these exercises could help your students be more reflexive in their future practice.

Future research directions

This book is the result of a single collaborative research project established by the Special Interest Group *Mathematics Birth to Eight Years* of the European Early Childhood Education Research Association. All researchers followed guidelines established by the SIG to generate the vignettes and commentaries which form the chapters of the book. In spite of the guidelines and the guidance from the editors of the book – or, perhaps, because of these – the chapters have their own character, reflecting the contexts of their origin and the beliefs and values of the researchers.

While it is not always clear, most of the children who feature in the vignettes are in the three to five years age range and display a great deal of mathematical knowledge and many skills. The educators and researchers who have interacted with the children as recorded in the vignettes are generally quite experienced and expert in their roles. The research recorded in this book sets a high standard.

Using Borton's (1970) notion of *Now what?* we can ask, *Now what is to be researched in early childhood mathematics education?* It would seem that the answer should lie in Bishop's universal mathematical activities and his emphasis on the importance of culture in mathematical learning. As our preschools become more diverse through the (sometimes forced) mobility of families and their children, there is an urgent need to research how children's mathematical learning can be facilitated in culturally unfamiliar circumstances. What scope is available to educators to investigate how the cultures of these immigrant, perhaps refugee, children impact on mathematics learning and to introduce some features of this into their classrooms?

Only two chapters in the book report a vignette which substantively uses digital technology to assist in mathematics learning. Many young children have access to such technology at home, and the education systems will want to stay abreast of the latest developments. This is a fertile field for researchers in mathematics education and beyond.

The notion of using a vignette to stimulate reflexivity in practising and prospective early childhood educators is not new but it is still innovative in many countries. For teacher education researchers in these countries this book offers a new way to generate and analyse data about teaching approaches.

One of the results of the research project reported in this book is an indication of just how innovative, reflexive, and resourceful early childhood educators and researchers can be. This bodes well for future research directions in early childhood mathematics education.

References

Bambino, D. (2002). Critical friends. *Educational Leadership, 59*(6), 25–27.

Bishop, A. J. (1988a). *Mathematical enculturation: A cultural perspective on mathematics education.* Dordrecht: Kluwer.

Bishop, A. J. (1988b). Mathematics education in its cultural context. *Educational Studies in Mathematics, 19*(2), 179–191.

Bishop, A. J. (2002). Critical challenges in researching cultural issues in mathematics education. *Journal of Intercultural Studies, 23*(2), 119–131. https://doi.org/10.1080/07256860220151041

Bolton, G. (2010). *Reflective practice: Writing and professional development* (3rd ed.). Los Angeles: SAGE.

Borton, T. (1970). *Reach, touch, and teach: Student concerns and process education.* New York: McGraw-Hill.

Clements, D. H., & Sarama, J. (2009). *Learning and teaching early math: The learning trajectories approach.* New York: Routledge.

Connors, K., & Seifer, S. D. (2005). Reflection in higher education service-learning. Retrieved from www.yumpu.com/en/document/read/26360619/reflection-in-higher-education-service-learning

Cushman, K. (1998). How friends can be critical as schools make essential changes. *Horace, 14*(5), 1–8.

Devlin, K. J. (1994). *Mathematics, the science of patterns: The search for order in life, mind, and the universe*. New York: Scientific American Library.

Devlin, K. J. (2012). *Introduction to mathematical thinking*. Palo Alto: Keith Devlin.

Dewey, J. (1916). *Democracy and education: An introduction to the philosophy of education*. New York: Macmillan.

Dewey, J. (1933). *How we think: A restatement of the relation of reflective thinking to the educative process*. Lexington, MA: DC Heath.

Dewey, J. (1934). *Art as experience*. New York: Putnam.

DfE (2017). *Statutory framework for the Early Years Foundation Stage*. Department for Education. Retrieved from www.gov.uk/government/uploads/system/uploads/attachment_data/file/596629/EYFS_STATUTORY_FRAMEWORK_2017.pdf.

Donaldson, M. (1978). *Children's minds*. Glasgow: Fontana/Collins.

Einstein, A. (1954). *Ideas and opinions*. New York: Crown Publishers.

Fuson, K. (1992). Research on whole number addition and subtraction. In D. A. Grouws (Ed.), *Handbook of research on mathematics teaching and learning: A project of the National Council of Teachers of Mathematics* (pp. 243–275). New York: Macmillan.

Galilei, G. (1623). *Il Saggiatore* [The assayer]. Rome: Giacomo Mascardi.

Karlsen, L., & Lekhal, R. (2019). Practitioner involvement and support in children's learning during free play in two Norwegian kindergartens. *Journal of Early Childhood Research, 17*(3), 233–246. https://doi.org/10.1177/1476718X19856390

Kolb, A., & Kolb, D. A. (2018). Eight important things to know about the Experiential Learning Cycle. *Australian Educational Leader, 40*(3), 8–14.

Kolb, A. Y., & Kolb, D. A. (2012). Experiential learning spiral. In N. M. Seel (Ed.), *Encyclopedia of the sciences of learning* (pp. 1212–1214). Boston, MA: Springer US.

Kolb, D. A. (2015). *Experiential learning. Experience as the source of learning and development* (2nd ed.). Upper Saddle River, NJ: Pearson Education.

Marcus, R. (2015). *Autonomy Platonism and the indispensability argument*. Lanham, MD: Rowman and Littlefield.

Ministry of Education and Research. (2017). *Framework plan for kindergartens – content and tasks*. Oslo: Ministry of Education and Research. Retrieved from www.udir.no/globalassets/filer/barnehage/rammeplan/framework-plan-for-kindergartens2-2017.pdf.

Nakken, A. H., & Thiel, O. (2019). *Matematikkens kjerne* [The core of mathematics] (2nd ed.). Bergen: Fagbokforlaget.

Nakken, A. H., Grimeland, Y., Nergård, B., & Thiel, O. (2016). *Young children's play in a mathematics room*. Poster presented at the 13th International Congress on Mathematics Education (ICME), Hamburg.

OECD. (2006). *Starting strong II: Early childhood education and care*. Paris: OECD.

Otsuka, K., & Jay, T. (2017). Understanding and supporting block play: Video observation research on preschoolers' block play to identify features associated with the development of abstract thinking. *Early Child Development and Care, 187*(5–6), 990–1003. https://doi.org/10.1080/03004430.2016.1234466

Piaget, J. (1936). *La naissance de l'intelligence chez l'enfant* [The origin of intelligence in the child]. Neuchâtel: Delachaux et Niestlé.

Poland, M., & van Oers, B. (2007). Effects of schematising on mathematical development. *European Early Childhood Education Research Journal, 15*(2), 269–293. https://doi.org/10.1080/13502930701321600.

Sæbbe, P.-E., & Samuelsson, I.P. (2017). Hvordan underviser barnehagelærere? Eller gjør man ikke det i barnehagen? [How do kindergarten teachers teach? Or don't they do that in kindergarten?] *Tidsskrift for Nordisk Barnehageforskning* [Journal of Nordic kindergarten research], *14*(7), 1–15. https://doi.org/10.7577/nbf.1731.

Sarama, J., & Clements, D. H. (2009). *Early childhood mathematics education research: Learning trajectories for young children.* New York: Routledge.

Schön, D. (1983). *The reflective practitioner.* New York: Basic Books.

Schön, D. (1987). *Educating the reflective practitioner.* San Francisco: Jossey-Bass.

Steen, L. A. (1988). The science of patterns. *Science, 240*(4852), 611–616.

Tuastad, S. E., Bjørnestad, E., & Alvestad, M. (2020). Contested quality: the struggle over quality, play and preschooling in Norwegian early childhood education and care. In S. Garvis & S. Phillipson (Eds.), *Policification of early childhood education and care* (pp. 154–174). London: Routledge. https://doi.org/10.4324/9780203730539.

van Hiele, P. M. (1957). *De problematiek van het inzicht : gedemonstreerd aan het inzicht van schoolkinderen in meetkunde-leerstof.* [The problem of insight, in connection with schoolchildren's insight into subject-matter of geometry] PhD dissertation, University Utrecht. Amsterdam: Meulenhoff.

van Hiele, P. M. (1986). *Structure and insight: A theory of mathematics education.* Orlando: Academic Press.

van Oers, B. (2001). Contextualisation for abstraction. *Cognitive Science Quarterly, 1*(3/4), 279–306.

White, M., & Gribbin, J. (1994). *Einstein: A life in science.* London: Simon & Schuster.

INDEX